The Integral:
A Crux for Analysis

Synthesis Lectures on Mathematics and Statistics

Editor
Steven G. Krantz, *Washington University, St. Louis*

The Integral: A Crux for Analysis

Steven G. Krantz

ISBN: 978-3-031-01273-0   paperback
ISBN: 978-3-031-02401-6   ebook

DOI 10.1007/978-3-031-02401-6

A Publication in the Springer series
SYNTHESIS LECTURES ON MATHEMATICS AND STATISTICS

Lecture #9
Series Editor: Steven G. Krantz, *Washington University, St. Louis*
Series ISSN
Synthesis Lectures on Mathematics and Statistics
Print 1938-1743   Electronic 1938-1751

The Integral:
A Crux for Analysis

Steven G. Krantz
Washington University, St. Louis

SYNTHESIS LECTURES ON MATHEMATICS AND STATISTICS #9

ABSTRACT

This book treats all of the most commonly used theories of the integral. After motivating the idea of integral, we devote a full chapter to the Riemann integral and the next to the Lebesgue integral. Another chapter compares and contrasts the two theories. The concluding chapter offers brief introductions to the Henstock integral, the Daniell integral, the Stieltjes integral, and other commonly used integrals.

The purpose of this book is to provide a quick but accurate (and detailed) introduction to all aspects of modern integration theory. It should be accessible to any student who has had calculus and some exposure to upper division mathematics.

KEYWORDS

integral, Riemann integral, Lebesgue integral, Henstock integral, Daniell integral, Stieltjes integral, limiting processes, inequalities, measure, measurable function, Lebesgue theorems

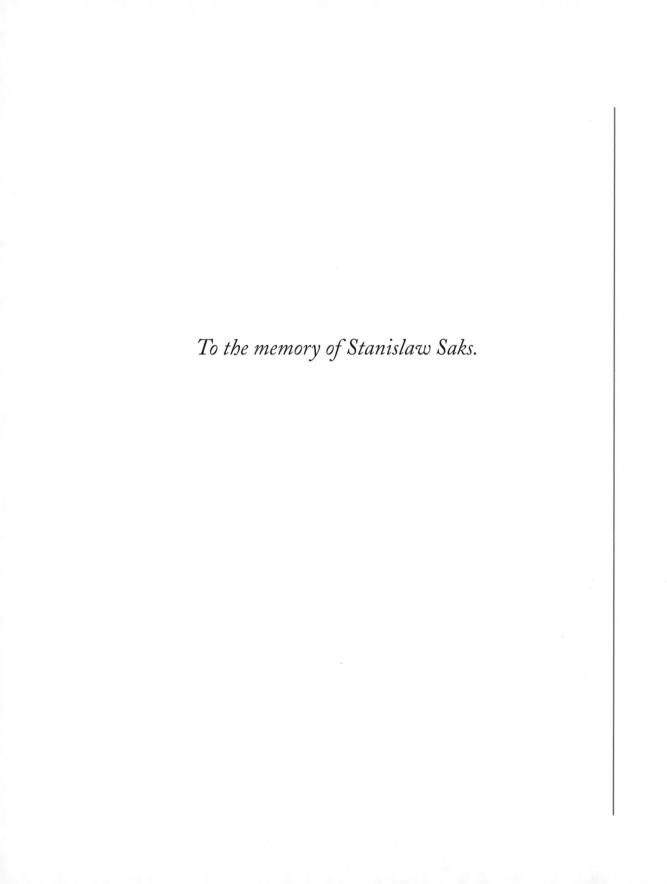

To the memory of Stanislaw Saks.

Contents

Preface

The integral is an old idea—a venerable part of the calculus. The integral is important because **(i)** it is dual to the derivative (as the Fundamental Theorem of Calculus tells us) and **(ii)** it is the most powerful and versatile tool in calculus. Most of real analysis is based on the theory of the integral. And the integral, in turn, inspires many of the key ideas of functional analysis.

There are many theories of the integral. One of the first modern theories was that enunciated by Bernhard Riemann in his celebrated paper on Fourier series. The Riemann integral is relatively simple, and it does the job nicely. It is certainly the most ubiquitous and widely-used integral in the world today. Mathematicians, physicists, engineers, and many others use the Riemann integral regularly.

But more advanced studies in analysis require an integral with better convergence properties. That is, if f_j is a sequence of functions on a bounded interval $[a, b]$, then we would like to know that

$$\lim_{j \to \infty} \int_a^b f_j(x)\, dx = \int_a^b \lim_{j \to \infty} f_j(x)\, dx\,. \tag{$*$}$$

For the Riemann integral, the standard condition to guarantee $(*)$ is uniform convergence of the sequence of functions $\{f_j\}$. In many analytical contexts, we would like to be able get away with a weaker and/or more flexible condition. We would also like a broader class of functions to be integrable. Enter the Lebesgue integral.

The Riemann integral is based on subdividing the domain of the function (by way of a partition). The Lebesgue integral is based on subdividing the range of the function. The latter operation requires measure theory, and that is a subtle idea. We cannot, in the present book, engage in a thoroughgoing treatment of measure theory. We instead give a quick-and-dirty introduction, adequate for a brisk development of the Lebesgue integral.

We develop these two fundamental integrals separately, each in its own chapter. But then we take some pains to compare and contrast the two integrals. Examples and specific theorems show the advantages of each.

We close the book by giving a quick survey of some of the other integrals that have been developed: the Henstock, the Daniell, the Stieltjes, and some others. In each case, we try to show how the integral being discussed is useful, and is different from the basic two.

We call this book *The Integral: A Crux for Analysis* as a nod to the central role of the integral in everything that we do. To be sure, the derivative is an important idea, but it is a relatively straightforward one. The integral, by contrast, is deep and profound. There is a lot to say about the idea of the integral, and we provide a pathway into the subject here.

This book is best read by a student who has had calculus and some exposure to upper division mathematics. The student need not have mastered undergraduate or graduate analysis. In fact, we would hope that this book would aid in that considerable task. The book may be read as an adjunct to any of the standard analysis texts, or it could be used as a freestanding text in a short course.

We hope that this book contributes to a broader understanding of fundamental principles of integration. This attribute should be useful for mathematical scientists and for many others as well.

Steven G. Krantz
January, 2011

CHAPTER 1

Introduction

1.1 WHAT IS AN INTEGRAL?

What is an integral? It is a form of addition, but it is addition of infinitely many quantities. It is different from a series. For a series deals with an infinite *discrete sum*, while an integral deals with an infinite *continuous sum*. Integrals are powerful tools for modeling physical processes (such as work, energy, volume, moments, etc.). They are also essential tools of mathematical analysis.

The theory of the integral teaches us that we can break a planar area up into strips or wedges and approximate the value of the area by the sum of the areas of the component pieces. Making the strips or wedges more narrow (and hence more numerous) makes the approximation more accurate. In the limit, the value becomes an integral, and it exactly equals the desired area. Likewise, one can approximate a spatial volume by a sum of wedges or columns. Summing the volumes of the component pieces gives an approximation to the volume of the original solid. Again, a limiting process shows that the actual, precise volume equals the value of an integral. Many other physical processes, such as work and moments of inertia, can be determined using integrals.

Because of the Fundamental Theorem of Calculus, we know that integrals are inextricably linked with derivatives. In fact, the two are inverse processes to one another. More will be said about this symbiosis later. It is the Fundamental Theorem that makes the integral a powerful tool in science and technology.

1.2 WHAT IS THE RIEMANN INTEGRAL?

In his fundamental paper [RIE], Bernhard Riemann introduced the Riemann integral almost as an afterthought. His aim in that paper was to prove some new results about Fourier and trigonometric series. And his point of view was that, to approach the matter successfully, one first had to have a clear idea of what the integral is. So he just stated what he needed.

The Riemann integral is based on the idea of dividing the domain of the function in question into small units. See Figure 1.1. Over each such unit, or subinterval, we erect an approximating rectangle. Refer to Figure 1.2. The sum of the areas of these rectangles approximates the area under the curve.

As the partition of the interval becomes finer, the number of subintervals becomes greater. And the approximating rectangles become narrower and more precise. Thus, the approximation to the area under the curve becomes more accurate. We let the lengths of the subintervals tend to zero, hence the value of the sum of the areas of the rectangles tends to the value of an integral. In the end, we declare the area under the curve to be equal to the value of the integral.

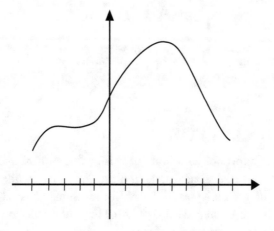

Figure 1.1: Dividing the domain into small units.

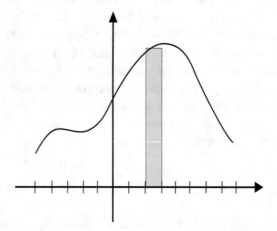

Figure 1.2: An approximating rectangle.

Although we motivate the concept of integral with area, in fact the integral is a freestanding mathematical concept that can be applied to a variety of different problems. A great many physical and mathematical quantities can be described with the integral.

1.3   WHAT IS THE RIEMANN INTEGRAL GOOD FOR?

Any physical or mathematical problem that entails aggregation, or addition, or summing of a quantity or process is amenable to analysis with an integral. [Even a series may be realized as a particular type of Riemann-Stieltjes integral.] Any mathematical operation that involves inverting a derivative can be addressed with an integral. As an instance, solving an ordinary or partial differential equation is often effected with an integral.

There are many means for evaluating integrals. Some of these techniques involve other mathematical subjects, such as complex analysis. Those integrals which cannot be evaluated explicitly are amenable to numerical techniques—generally implemented with the aid of a computer (see Section 2.5 below). These numerical techniques may be applied to achieve any desired degree of accuracy.

1.4   WHAT IS THE RIEMANN INTEGRAL NOT GOOD FOR?

The Riemann integral has certain built-in limitations. It can only be applied to a fairly restrictive class of functions. We shall actually give a complete characterization of those functions which can be integrated in the sense of Riemann. Perhaps more significantly, there are restrictions on which limiting processes the Riemann integral respects. Since many sophisticated ideas in mathematical analysis entail commuting limits, this restriction is fundamental and, in practice, confining.

Other theories of the integral were developed in order to address some of the shortcomings of the Riemann integral. The most prominent of these is the Lebesgue integral. A good portion of this book will be devoted to the Lebesgue integral.

1.5   WHAT IS THE LEBESGUE INTEGRAL?

The Lebesgue integral is founded on Henri Lebesgue's (1875–1941) theory of measure. The idea of measure theory is that we want to assign a length to each subset of the real numbers. Unfortunately, this is impossible to do in a logically consistent fashion. So measure theory tells us how to pick out which sets we can measure and how to measure them. We call such sets measurable. If A is a measurable set, then its measure or length is denoted by $m(A)$.

The fundamental building block of the Lebesgue theory is the characteristic function. Let A be a set. Then the characteristic function of A is

$$\chi_A(x) = \begin{cases} 1 & \text{if} & x \in A \\ 0 & \text{if} & x \notin A. \end{cases}$$

Then the Lebesgue integral of χ_A is

$$\int \chi_A(x)\, dx = m(A).$$

If f is an arbitrary nonnegative function on the real line, then we can approximate f by a sum of constants times characteristic functions (i.e., a linear combination of characteristic functions). So we approximate the integral of f by the sum of those constants times the measures of those sets.

Although Henri Lebesgue is usually credited with the creation of the integral that bears his name, there is evidence that Leonida Tonelli (1885–1946) developed some of the same ideas at about the same time. The book [ROY] gives background and a thorough treatment of the Lebesgue integral.

1.6   WHAT IS THE LEBESGUE INTEGRAL GOOD FOR?

The Lebesgue integral is good for doing mathematical analysis. With this integral, the collection of functions that can be integrated is quite large.

Perhaps more significantly, the limiting processes that the Lebesgue integral respects are broad-ranging and flexible. So significant theorems about the convergence of Fourier series, about singular integrals, about Fourier integral operators, and about other important artifacts of analysis can be attacked using the Lebesgue integral.

1.7   WHAT IS THE LEBESGUE NOT GOOD FOR?

The Lebesgue integral is a powerful and flexible tool, but it is not a panacea. There are many important questions of analysis—such as the convergence of multidimensional Fourier series—that we cannot understand using the tools at hand. It is possible that a new theory of the integral will be needed to make any progress.

EXERCISES

1. Let S be the unit circle in the plane. Inscribe an equilateral triangle in S. What is the area inside that triangle? Now inscribe a square in the circle S. What is the area inside that square? Do the same with a regular pentagon and a regular hexagon. Now calculate the area inside a regular j-gon inscribed in S. What is the limit of these areas as $j \to \infty$?

2. Give an example of a sequence of nonnegative, continuous functions $f_j : \mathbb{R} \to \mathbb{R}$, each bounded by 1, so that **(i)** the area above the x-axis and under the graph of f_j is 1 for each j and **(ii)** the pointwise limit $\lim_{j \to \infty} f_j(x)$ is identically 0. What does this example suggest about limiting operations and integration?

3. The Cantor ternary set C has zero length but uncountably many elements—see [KRA1]. Thus, C is small by one measure but large by another measure. Imitate Cantor's construction to produce a Cantor-like set that has uncountably many elements but positive length.

4. Imitate the scheme in Exercise 1 to estimate the circumference of the circle S. Do so by calculating the perimeter of the inscribed triangle, then the perimeter of the inscribed square, and so forth. What is the limiting value as $j \to \infty$?

5. Modify the example in Exercise 2 as follows. Let f_j be nonnegative continuous functions so that **(i)** Each f_j is supported in the interval $[0, 1/j]$ (i.e., f_j is identically zero outside that interval) and **(ii)** Each f_j has integral 1. Analyze this sequence and say what it tells us about limiting operations and integration.

6. Define
$$f_j(x) = \begin{cases} 2^j & \text{if} \quad j/2^j \le x \le (j+1)/2^j \\ -2^j & \text{if} \quad (j+1)/2^j \le x \le (j+2)/2^j . \end{cases}$$

What does this sequence tell us about limiting operations and integration?

7. Define

$$f_j(x) = \sin(jx).$$

What does this sequence tell us about limiting operations and integration?

C H A P T E R   2

The Riemann Integral

2.1   THE DEFINITION

Let f be a real-valued function with domain the interval $[a, b]$. See Figure 2.1. A *partition* of $[a, b]$ is a list of points $\mathcal{P} \equiv \{x_0, x_1, x_2, \ldots, x_k\}$ with

$$a = x_0 < x_1 < x_2 < \cdots < x_{k-1} < x_k = b\,.$$

See Figure 2.2. In many contexts, it is useful to assume that $\triangle_j \equiv x_j - x_{j-1}$ is a constant $\triangle x$ independent of $j = 1, 2, \ldots, k$. But greater generality, and greater flexibility in the proofs, is attained if we allow the partition to be more arbitrary.

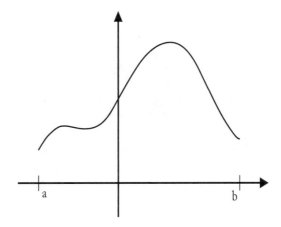

Figure 2.1: A continuous function f on $[a, b]$.

In each interval $I_j = [x_{j-1}, x_j]$, we select a point ξ_j. Then we define the *Riemann sum*

$$\mathcal{R}_\mathcal{P} \equiv \sum_{j=1}^{k} f(\xi_j) \, \triangle x_j\,.$$

A typical Riemann sum is depicted in Figure 2.3. Clearly, the Riemann sum $\mathcal{R}_\mathcal{P}$ depends on the choice of the ξ_j, but we do not indicate that dependence explicitly with the notation.

Define the *mesh* of the partition $\mathcal{P}$ to be

$$m(\mathcal{P}) \equiv \max_{j} (x_j - x_{j-1})\,.$$

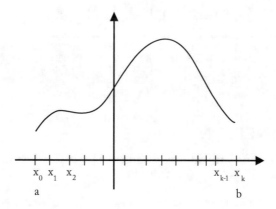

Figure 2.2: A partition.

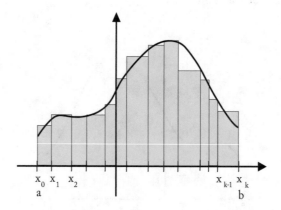

Figure 2.3: A Riemann sum.

We say that the Riemann sums have a limit ℓ as the mesh of the partitions tends to zero if, given $\epsilon > 0$, there is a $\delta > 0$ such that, if $m(\mathcal{P}) < \delta$, then $|\mathcal{R}_\mathcal{P} - \ell| < \epsilon$. In that case, we call the limit ℓ the *Riemann integral* of the function f on the interval $[a, b]$. We write

$$\int_a^b f(x)\,dx$$

for the Riemann integral.

The good news is that, for a continuous function f on a compact interval $[a, b]$, the integral always exists:

Theorem 2.1 *Let f be a continuous function on the bounded interval $[a, b]$. Then the Riemann integral*

$$\int_a^b f(x) \, dx$$

exists and is finite.

Proof. Of course f is uniformly continuous on $[a, b]$. Let $\epsilon > 0$. Choose a $\delta > 0$ such that $|s - t| < \delta$ implies $|f(s) - f(t)| < \epsilon$.

Now let $\mathcal{P}$ and $\mathcal{Q}$ be partitions of $[a, b]$ with mesh less than δ. We want to compare $\mathcal{R}_{\mathcal{P}}$ with $\mathcal{R}_{\mathcal{Q}}$. We do so by comparing each of them with the Riemann sum arising from the *common refinement* of $\mathcal{P}$ and $\mathcal{Q}$. The latter is defined to be

$$\mathcal{W} \equiv \mathcal{P} \cup \mathcal{Q}.$$

Let the points of $\mathcal{P}$ be called $p_0, p_1, \ldots, p_k$ and let the points of $\mathcal{W}$ be called $w_0, w_1, \ldots, w_m$. The partition $\mathcal{P}$ gives rise to subintervals I_j, having lengths $\triangle_j$, and the partition $\mathcal{W}$ gives rise to subintervals J_ℓ, having lengths $\triangle'_\ell$. For each j, let ξ_j as usual be a point chosen from I_j and for each ℓ, let μ_ℓ be a point chosen from J_ℓ (these points are chosen to enable us to write down the Riemann sums). Then we have

$$
\begin{aligned}
|\mathcal{R}_P - \mathcal{R}_{\mathcal{W}}| &= \left| \sum_{j=1}^k f(\xi_j) \, \triangle_j - \sum_{\ell=1}^m f(\mu_\ell) \triangle'_\ell \right| \\
&= \left| \sum_{j=1}^k f(\xi_j) \, \triangle_j - \sum_{j=1}^k \left[\sum_{J_\ell \subseteq I_j} f(\mu_\ell) \triangle'_\ell \right] \right| \\
&= \left| \sum_{j=1}^k \left[f(\xi_j) \sum_{J_\ell \subseteq I_j} \triangle'_\ell \right] - \sum_{j=1}^k \left[\sum_{J_\ell \subseteq I_j} f(\mu_\ell) \triangle'_\ell \right] \right| \\
&\leq \sum_{j=1}^k \sum_{J_\ell \subseteq I_j} \left| f(\xi_j) - f(\mu_\ell) \right| \triangle'_\ell \\
&\leq \sum_{j=1}^k \sum_{J_\ell \subseteq I_j} \epsilon \, \triangle'_\ell \\
&\leq (b - a) \cdot \epsilon.
\end{aligned}
$$

So we see that $|\mathcal{R}_{\mathcal{P}} - \mathcal{R}_{\mathcal{W}}| \leq (b - a) \cdot \epsilon$ and an identical argument shows that $|\mathcal{R}_{\mathcal{Q}} - \mathcal{R}_{\mathcal{W}}| \leq (b - a) \cdot \epsilon$. It follows that $|\mathcal{R}_{\mathcal{P}} - \mathcal{R}_{\mathcal{Q}}| \leq 2(b - a) \cdot \epsilon$. Thus, we see that the Riemann sums are Cauchy, and they have a finite limit ℓ. $\square$

2.2 PROPERTIES OF THE RIEMANN INTEGRAL

Now the most important property of the Riemann integral is its convergence feature. First, we need a lemma.

Lemma 2.2   *If f is a continuous function on the bounded interval $[a, b]$ then*

$$\left| \int_a^b f(x)\,dx \right| \le \int_a^b |f(x)|\,dx .$$

Proof. Let $\mathcal{P}$ be a partition of $[a, b]$ and let $\mathcal{R}_\mathcal{P}$ be a corresponding Riemann sum. Then

$$|\mathcal{R}_P| = \left| \sum_{j=1}^k f(\xi_j)\triangle_j \right| \le \sum_{j=1}^k |f(\xi_j)|\,\triangle_j .$$

Now letting the mesh of $\mathcal{P}$ tend to zero gives the result. □

Theorem 2.3   *Let f_j be continuous functions on a bounded interval $[a, b]$ and suppose that the f_j converge uniformly on $[a, b]$ to a limit function f. Then*

$$\lim_{j \to \infty} \int_a^b f_j(x)\,dx = \int_a^b f(x)\,dx .$$

Proof. Let $\epsilon > 0$. Choose N so large that if $j > N$, then $|f_j(x) - f(x)| < \epsilon$ for all $x \in [a, b]$. Let $\mathcal{P} = \{p_0, p_1, \ldots, p_k\}$ be a partition of $[a, b]$. In each interval $[p_{\ell-1}, p_\ell]$, choose a point ξ_ℓ as usual. Let $\mathcal{R}_\mathcal{P}$ be the corresponding Riemann sum for the function f and $\mathcal{R}_\mathcal{P}^j$ the Riemann sum for f_j. Then

$$\begin{aligned}
|\mathcal{R}_\mathcal{P} - \mathcal{R}_\mathcal{P}^j| &= \left| \sum_{\ell=1}^k f(\xi_\ell)\,\triangle_\ell - \sum_{\ell=1}^k f_j(\xi_\ell)\triangle_\ell \right| \\
&\le \sum_{\ell=1}^k |f(\xi_\ell) - f_j(\xi_\ell)|\triangle_\ell < \sum_{\ell=1}^k \epsilon\,\triangle_\ell \\
&= (b-a) \cdot \epsilon .
\end{aligned}$$

Letting the mesh of $\mathcal{P}$ tend to zero now gives

$$\left| \int_a^b f(x)\,dx - \int_a^b f_j(x)\,dx \right| \le (b-a) \cdot \epsilon .$$

This is the convergence result that we want. □

Of course the Riemann integral has a number of basic properties, many of them intuitively appealing, that we must record. We do so now.

Proposition 2.4   *The Riemann integral is a linear operator. That is to say, if f, g are continuous functions on the bounded interval* $[a, b]$ *and if* α, β *are real scalars, then*

$$\int_a^b \alpha f(x) + \beta g(x)\, dx = \alpha \int_a^b f(x)\, dx + \beta \int_a^b g(x)\, dx\,.$$

Proof. Clearly, any given Riemann sum is a linear operator on the function, which is the argument of the sum. The result then follows by passing to the limit.   □

Remark 2.5   It is certainly possible, and very natural, to consider both Riemann and Lebesgue integration of complex-valued functions. However, to keep things simple, we shall concentrate in this book on real-valued functions.

Proposition 2.6   *Let* $[a, b]$ *be a nontrivial, bounded interval in the real line, and let f be a continuous function on* $[a, b]$. *Let* $a < c < b$. *Then*

$$\int_a^c f(x)\, dx + \int_c^b f(x)\, dx = \int_a^b f(x)\, dx\,. \tag{$\star$}$$

Proof. Let $\mathcal{P}$ be a partition of $[a, b]$. Suppose that c lies in the subinterval I_j. For simplicity, we suppose for the moment that $x_{j-1} < c < x_j$. Then

$$a = x_0 < x_1 < \cdots < x_{j-1} < c < x_j < x_{j+1} < \cdots < x_k$$

and we see that $\mathcal{P}_1 \equiv \{x_0, x_1, \ldots, x_{j-1}, c\}$ is a partition of $[a, c]$, and $\mathcal{P}_2 \equiv \{c, x_j, x_{j+1}, \ldots, x_k\}$ is a partition of $[c, b]$. Then

$$\mathcal{R}_{\mathcal{P}} f = \mathcal{R}_{\mathcal{P}_1} f + \mathcal{R}_{\mathcal{P}_2} f\,.$$

Taking the limit, as the mesh of the partition tends to zero, the equality $(\star)$ follows.

It is easy to see that, if $c = x_{j-1}$ or $c = x_j$, then a slight modification of this argument will yield the same conclusion.   □

Remark 2.7   We see, in this last proof, that it is useful to be able to consider partitions in which not all the subintervals have equal length. Even if $\mathcal{P}$ has subintervals of equal length, the sub-partitions $\mathcal{P}_1$ and $\mathcal{P}_2$ will not.

In case $b < a$, then we define

$$\int_a^b f(x)\,dx \equiv -\int_b^a f(x)\,dx\,.$$

With this definition, we see that all the properties of the Riemann integral that we treated in this section still hold when the lower limit of integration exceeds the upper limit. In particular, Proposition 2.6 on dividing the domain of integration still applies.

2.3   CHARACTERIZATION OF RIEMANN INTEGRABILITY

There is a lovely classical result that gives necessary and sufficient conditions for a function to be Riemann integrable. We discuss that idea now. The proof that we present is taken from [BRO].

We first need to consider the ancillary idea of outer measure. Say that $S \subseteq \mathcal{R}$ is a subset. A collection $\{I_j\}$ of open intervals is called an *open covering* of S if $\cup_j I_j \supseteq S$. Then we set

$$o(S) = \inf \sum_j |I_j|\,,$$

where the infimum is taken over all possible open coverings of S. Here $|I_j|$ denotes the length of the interval I_j.

Outer measure is a subtle idea. For example, we have the following result:

Lemma 2.8   *Let $S \subseteq \mathcal{R}$ be a countable set. Then S has outer measure 0.*

Proof. Let $S = \{s_1, s_2, \dots\}$. Let $\epsilon > 0$. For each $j = 1, 2, \dots$, let

$$I_j = (s_j - \epsilon/2^{j+1}, s_j + \epsilon/2^{j+1})\,.$$

Then clearly

$$\bigcup_j I_j \supseteq S\,.$$

Furthermore,

$$\sum_j |I_j| = \sum_j 2^{-j}\epsilon = \epsilon\,.$$

Since this is true for any $\epsilon > 0$, we conclude that $o(S) = 0$.   □

Definition 2.9   Let f be a function on an interval (c, d). Let $x_0 \in (c, d)$. We define the *oscillation* of f at x_0 to be

$$\operatorname{osc} f(x_0) = \lim_{\delta \to 0} \sup_{\substack{|x_1 - x_0| < \delta \\ |x_2 - x_0| < \delta}} |f(x_1) - f(x_2)|.$$

Clearly, the function f is continuous at x_0 if and only if $\operatorname{osc} f(x_0) = 0$.

Theorem 2.10   (Lebesgue)   *A function f on a bounded interval $[a, b]$ is Riemann integrable if and only if*

(a) *f is bounded;*

(b) *The set of discontinuities of f has outer measure 0.*

Proof. First, suppose that f is Riemann integrable. It is clear that f must be bounded because if not, and if $|f(t_j)| \to +\infty$, then we may arrange for Riemann sums to be arbitrarily large (just by choosing the ξ_j from among the t_j). So they will not converge, and the function will not be Riemann integrable.

So we may suppose that f is bounded. Let T denote the set of points at which f is discontinuous. Seeking a contradiction, we suppose that T has positive outer measure. We can write

$$T = \bigcup_{j=-\infty}^{\infty} \{x \in T : 2^{-(j+1)} \leq \operatorname{osc} f(x) < 2^{-j}\}.$$

Since T has positive outer measure, one of the subsets on the right will have positive outer measure. Thus, there will be a number $\alpha > 0$ and a subset $T_1 \subseteq T$ of outer measure $\tau > 0$ so that f has oscillation at least α at each point of T_1.

Now let $\mathcal{P}$ be any partition of $[a, b]$. Those subintervals in the partition that contain in their interiors points of T_1 must have lengths totaling at least τ. Any interval containing a point of T_1 will, in fact, have two points x_1 and x_2 in it, arbitrarily close together, (by definition of oscillation) so that $|f(x_1) - f(x_2)| > \alpha/2$. In fact, we may always suppose that $f(x_2) > f(x_1)$. If $\mathcal{Q} = \{q_1, q_2, \ldots, q_k\}$ is any partition of $[a, b]$, with no matter how fine a mesh, then we can arrange in each subinterval that contains an element of T_1 that there are x_1 and x_2 as above. Thus, we may consider two different Riemann sums $\mathcal{R}_1$ and $\mathcal{R}_2$, modeled on $\mathcal{Q}$, which agree in their choice of ξ_j in any subinterval that does not contain points of T_1 in its interior but for which we choose x_1 (resp. x_2) in an interval that does contain points of T_1 in its interior. Then

$$|\mathcal{R}_2 - \mathcal{R}_1| \geq \tau \cdot \frac{\alpha}{2}.$$

Since this inequality holds no matter how small the mesh of the partition, we see that the Riemann integral cannot exist. Hence, T cannot have positive outer measure.

Conversely, let us suppose that $|f|$ is bounded by a positive number M in $[a, b]$ and that T (as defined above) has outer measure zero. We shall show that f must be Riemann integrable on $[a, b]$. We claim that, for any $\eta > 0$, there is a $\delta > 0$ so that, if S_1 is the Riemann sum $\sum_{j=1}^{n_1} f(x_j) \triangle x_j$ and S_2 is the Riemann sum $\sum_{j=1}^{n_2} f(x'_j) \triangle' x_j$, with $\triangle x_j < \delta$ and $\triangle' x_j < \delta$ for all j, then $|S_1 - S_2| < \eta$. To see this, first note that T can be covered by open intervals the sum of whose lengths is less than $\eta/(8M)$. Let A be the union of those intervals. Let $B = [a, b] \setminus A$. Clearly, B is closed.

Let $\delta > 0$ be chosen so that, if $x_0 \in B$ and $x \in [a, b]$ with $|x - x_0| < \delta$, then $|f(x) - f(x_0)| < \eta/[8(b - a)]$. This is a slight variant of the uniform continuity property for continuous functions on a compact set.

Now consider the partition $\mathcal{W} = \{w_0, w_1, \ldots, w_{n_3}\}$, which is the common refinement of the two considered in the last paragraph but one. Let the increments in this partition be denoted by $\triangle''_j$. Set

$$S_3 = \sum_{j=1}^{n_3} f(\xi_j) \triangle''_j .$$

We may, in each instance, take ξ_j to be the midpoint of the corresponding subinterval.

Now we may calculate that

$$|S_1 - S_3| \le \sum_{j=1}^{n_1} |f(x_j) - f(w'_j)| \triangle x_j ,$$

where each w'_j is chosen from the jth subinterval of the partition for S_1 so as to maximize $|f(x_j) - f(w'_j)|$.

There are two types of terms to consider in this last sum. The subintervals for S_1 containing a point of B contribute to the sum less than

$$\frac{\eta}{4(b - a)} \cdot (b - a) = \frac{\eta}{4} .$$

Those intervals containing no point of B are, therefore, subsets of A and hence contribute less than

$$\frac{2M \cdot \eta}{8M} = \frac{\eta}{4} .$$

Thus,

$$|S_1 - S_3| < \frac{\eta}{4} + \frac{\eta}{4} = \frac{\eta}{2} .$$

A similar argument shows that

$$|S_2 - S_3| < \frac{\eta}{2} .$$

We may conclude therefore that

$$|S_1 - S_2| < \eta .$$

This proves the claim.

The first conclusion we may draw from our claim is that, if we set

$$\mathcal{S}_p = \sum_{j=1}^{p} f(\xi_j) \cdot \frac{1}{p} \,,$$

with $\xi_j = a + j(b-a)/p$, then the sequence $\mathcal{S}_1, \mathcal{S}_2, \ldots$ converges to a finite limit R.

Now, given $\epsilon > 0$, take $\delta > 0$ to correspond to $\eta = \epsilon/2$ in our claim above. Also, take $N > (b-a)/\delta$ so large that $|R - \mathcal{S}_N| < \epsilon/2$. Now, for any partition of $[a,b]$ into n intervals of length less than δ, with ξ_j an arbitrarily chosen point in the jth subinterval for each j, we let $S_1 = \sum_{j=1}^{n} f(\xi_j) \triangle x_j$ and $S_2 = \mathcal{S}_N$. Then

$$|S_1 - S_N| = |S_1 - S_2| < \eta = \frac{\epsilon}{2} \,.$$

Hence, $|R - S_1| < \epsilon$. It follows that f is Riemann integrable on $[a,b]$. $\square$

2.4   THE FUNDAMENTAL THEOREM OF CALCULUS

A simple statement of the Fundamental Theorem is as follows:

Theorem 2.11   *Let f be a continuous function on the bounded interval $[a,b]$. For $x \in [a,b]$, set*

$$F(x) = \int_a^x f(t) \, dt \,.$$

Then, for $x \in (a,b)$,

$$F'(x) = f(x) \,.$$

Proof. Let $\epsilon > 0$. By uniform continuity, choose a $\delta > 0$ so that $|s - t| < \delta$ implies $|f(s) - f(t)| < \epsilon$. Now we calculate, for $|h| < \delta$, that

$$
\begin{aligned}
\frac{F(x+h) - F(x)}{h} &= \frac{\int_a^{x+h} f(t)\,dt - \int_a^x f(t)\,dt}{h} \\
&= \frac{\int_x^{x+h} f(t)\,dt}{h} \\
&\leq \frac{\int_x^{x+h} f(x) + \epsilon \, dt}{h} \\
&= f(x) + \epsilon \,.
\end{aligned}
$$

A similar calculation shows that

$$\frac{F(x+h) - F(x)}{h} \geq f(x) - \epsilon.$$

Since this is true for any $\epsilon > 0$, we obtain the desired conclusion. $\qquad\square$

2.5   NUMERICAL TECHNIQUES OF INTEGRATION

2.5.1   INTRODUCTION

Many of the most important integrals in mathematics, such as

$$\mathrm{erf}(N) = \int_{-\infty}^{N} e^{-x^2} \, dx$$

(which represents the Gaussian normal distribution of probability theory), cannot be evaluated by standard methods of calculus. The reason is that it is impossible to write down an antiderivative for e^{-x^2} with a closed formula involving functions that we know (see [RIT] for a characterization of those functions that have closed form integrals). Nevertheless, it is crucial to be able to calculate such integrals to any desired degree of accuracy. The purpose of this section is to acquaint you with some techniques for doing so.

2.5.2   THE METHOD OF RECTANGLES

Recall that the *definition* of the integral was as a limit of Riemann sums. Let us agree that, throughout this section, we will only use Riemann sums over uniform partitions (partitions having intervals of equal length): $\Delta x_j = \Delta x = (b-a)/k$ for all j. We will also take $\xi_j \in I_j = [x_{j-1}, x_j]$ to be the right endpoint x_j. Then

$$\int_a^b f(x) \, dx = \lim_{k \to \infty} \sum_{j=1}^{k} f(x_j) \cdot \Delta x$$

where, as the limit is being taken, the mesh of the partition $\{x_1, ..., x_k\}$,

$$\Delta x = x_j - x_{j-1} = \frac{b-a}{k},$$

tends to zero. Refer to Figure 2.4.

By the definition of limit, the value of the Riemann sum is close to the value of the integral when the mesh is small. Again referring to Figure 2.4, we interpret this statement to mean that the value of the integral is closely approximated by the sum of the areas of the rectangles in the figure.

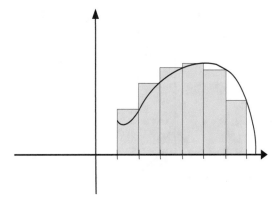

Figure 2.4: Approximation of the integral by a Riemann sum.

We summarize all these ideas:

Theorem 2.12   (The Method of Rectangles)   *Let f be a continuous function whose domain contains the interval $[a, b]$. The value of the integral*

$$\int_a^b f(x)\, dx$$

is approximated, using the uniform partition $\{x_0, \ldots, x_k\}$, by

$$f(x_1)\Delta x + f(x_2)\Delta x + \cdots + f(x_k)\Delta x.$$

Example 2.13   Use a Riemann sum with mesh $1/4$ to approximate the value of

$$\int_0^1 \frac{1}{1+x^2}.$$

Referring to Figure 2.5, we see that we must add up the areas of four rectangles. We have, with $f(x) = 1/(1+x^2)$,

$$
\begin{aligned}
\int_0^1 \frac{1}{1+x^2}\, dx \;\; &\approx \;\; f(1/4)\cdot\frac{1}{4} + f(1/2)\cdot\frac{1}{4} + f(3/4)\cdot\frac{1}{4} + f(1)\cdot\frac{1}{4} \\
&= \;\; \frac{16}{17}\cdot\frac{1}{4} + \frac{4}{5}\cdot\frac{1}{4} + \frac{16}{25}\cdot\frac{1}{4} + \frac{1}{2}\cdot\frac{1}{4} \\
m \;\; &= \;\; \frac{2449}{3400} \\
&\approx \;\; 0.72029412.
\end{aligned}
$$

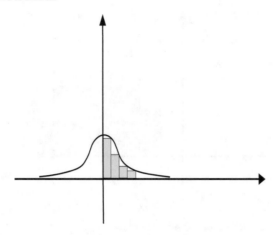

Figure 2.5: A Riemann sum for $f(x) = 1/(1 + x^2)$.

Now for our first example, we have chosen an integral which actually *can* be calculated in closed form. Indeed,

$$\int_0^1 \frac{1}{1 + x^2} \, dx = \text{Tan}^{-1} x \Big|_0^1 = \frac{\pi}{4} \approx 0.78539816.$$

So we see that our approximation is accurate to within 0.07.

Remark 2.14   The all-important issue of accuracy must always be considered. In Example 2.13, we can see how accurate our approximation is because we have another means at our disposal for evaluating the integral. But numerical methods are of greatest value when they can be applied in instances where no alternative is available.

Error estimates are studied in detail in a numerical analysis course. It can be shown that, if $|f'(x)| \leq M$ on the interval $[a, b]$ of integration, then the error in applying the method of rectangles will not exceed

$$\frac{M}{2} \cdot \frac{(b - a)^2}{k},$$

where k is the number of intervals in the partition. For instance, in Example 2.13, $f'(x) = -2x/(1 + x^2)^2$, which in absolute value is bounded by $M = 2$ for x in the interval $[0, 1]$. Since $b - a = 1$ and $k = 4$, we could have anticipated that our error would not exceed 0.25. In fact, we did much better than that.

Example 2.15   Determine how fine a partition would be required to estimate

$$\int_0^1 e^{-x^2} \, dx,$$

with a Riemann sum, to an accuracy of two decimal places.

We take $f(x) = e^{-x^2}$ so that

$$f'(x) = -2x \cdot e^{-x^2}$$

which certainly does not, in absolute value, exceed 2 on the interval $[0, 1]$ (we use here the simple observation that e^{-x^2} is never greater than 1). To achieve the desired accuracy, we need that

$$\frac{M \cdot (b-a)^2}{2k} < 5 \cdot 10^{-3}$$

or

$$\frac{2 \cdot 1}{2k} < 5 \cdot 10^{-3}.$$

We see that $k = 201$ will suffice for the desired degree of accuracy.

Example 2.13 illustrates rather vividly the attractiveness of a high speed computer for doing numerical integration. Performing 6401 calculations, even easy ones, is a tedious task. A mathematician's point of view is to try to find a better algorithm that will achieve excellent accuracy with many fewer calculations. We now consider some more efficient algorithms.

2.5.3   THE TRAPEZOIDAL RULE

The trouble with approximating by rectangles is that the error contributed by each rectangle is roughly a triangle (Figure 2.6) whose area is about half the square of the base of the rectangle.

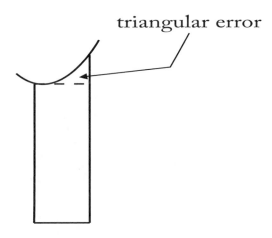

Figure 2.6: The error is roughly a triangle.

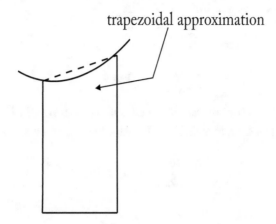

Figure 2.7: Replacing the rectangle by a trapezoid.

We need a method with the property that the error contribution from each piece is much smaller. Think geometrically: why not replace the rectangle by a trapezoid as in Figure 2.7?

The error certainly appears to be much smaller; in fact, if $|f''(x)|$ does not exceed M, then the accuracy of approximating the integral by this technique can be determined not to exceed

$$\frac{M \cdot (b - a)^3}{12k^2}.$$

But how do we calculate with the trapezoidal algorithm? Examine Figure 2.8.

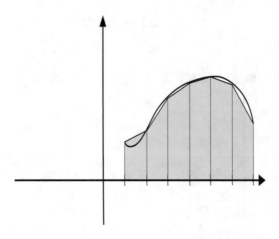

Figure 2.8: Calculating with the trapezoidal algorithm.

The partition of the interval $[a, b]$ is

$$a = x_0 < \cdots < x_k = b.$$

The sum of the areas of the trapezoids is

$$\frac{1}{2} \cdot \{f(x_0) + f(x_1)\} \cdot \Delta x + \frac{1}{2} \cdot \{f(x_1) + f(x_2)\} \cdot \Delta x + \cdots + \frac{1}{2} \cdot \{f(x_{k-1}) + f(x_k)\} \cdot \Delta x$$

$$= \frac{\Delta x}{2} \cdot \{f(x_0) + 2f(x_1) + 2f(x_2) + \cdots + 2f(x_{k-1}) + f(x_k)\}.$$

Let us summarize:

Theorem 2.16   (The Trapezoidal Rule)   *Let f be a continuous function whose domain contains the interval $[a, b]$. The value of the integral*

$$\int_a^b f(x)\,dx$$

is approximated, using the equally spaced partition $\{x_0, x_1, \ldots, x_k\}$, by the quantity

$$\frac{\Delta x}{2} \{f(x_0) + 2f(x_1) + 2f(x_2) + \cdots + 2f(x_{k-1}) + f(x_k)\}.$$

If $|f''(x)| \leq M$ on $[a, b]$ then the approximation is accurate to within

$$\frac{M \cdot (b - a)^3}{12k^2}.$$

Let us now redo Example 2.13 using the trapezoidal rule.

Example 2.17   Estimate

$$\int_0^1 \frac{1}{1 + x^2}\,dx$$

using the trapezoidal rule with $k = 4$.

With $f(x) = 1/(1 + x^2)$, we calculate that

$$f''(x) = \frac{6x^2 - 2}{(1 + x^2)^3}$$

hence $|f''(x)| \leq |6x^2 - 2| \leq 4$. Therefore, we know in advance that our error will not exceed

$$\frac{M \cdot (b - a)^3}{12k^2} = \frac{4 \cdot 1^3}{12 \cdot 4^2} \approx 0.02083333.$$

Now the trapezoidal rule says that the integral is approximately equal to

$$\frac{0.25}{2} \cdot \left\{ 1 + 2 \cdot \frac{16}{17} + 2 \cdot \frac{4}{5} + 2 \cdot \frac{16}{25} + \frac{1}{2} \right\} = \frac{1}{8} \cdot \frac{5323}{850} \approx 0.78279411.$$

Referring to Example 2.13, we see that the true value of the integral is

$$0.78539816\ldots$$

and our estimate does indeed fall within the anticipated range of accuracy. In addition, we achieved much greater accuracy than with the method of rectangles in Example 2.13, and with practically no extra effort.

Example 2.18   Use the trapezoidal rule to estimate

$$\int_0^1 e^{-x^2}\, dx$$

to two decimal place accuracy.

With $f(x) = e^{-x^2}$, we have

$$f''(x) = (4x^2 - 2)e^{-x^2}.$$

Therefore, $|f''(x)| \leq 2$ when $x \in [0, 1]$ (because the expression in parentheses cannot be larger than 2 and the exponential cannot be larger than 1). Therefore, the error in estimating the integral by the trapezoidal rule with a mesh of k equal subintervals is

$$\frac{2 \cdot (1 - 0)^3}{12 \cdot k^2}.$$

This quantity will be less than $5 \cdot 10^{-3}$ when $k > 5$. So we take $k = 6$, and our approximation by the trapezoidal rule is

$$\frac{1/6}{2} \cdot \left\{ 1 + 2 \cdot e^{-1/36} + 2 \cdot e^{-1/9} + 2 \cdot e^{-1/4} + 2 \cdot e^{-4/9} + 2 \cdot e^{-25/36} + e^{-1} \right\}$$
$$= 0.745119412\ldots.$$

Because the error is smaller than $5 \cdot 10^{-3}$, we know that the value of the integral is 0.75, accurate to within two decimal places.

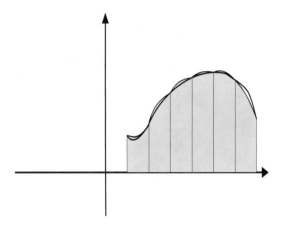

Figure 2.9: Replacing trapezoids by parabolas.

2.5.4   SIMPSON'S RULE

The technique for numerical calculation of integrals which is used most often in practice is Simpson's Rule. It builds on what we have done so far by approximating f *not* with rectangles, *not* with trapezoids, but with parabolas. Look at Figure 2.9.

In order to derive a formula for Simpson's Rule, we first need to do a calculation of the area under a parabola:

Lemma 2.19   *If*

$$P(x) = Ax^2 + Bx + C$$

is a parabola (either upward or downward opening) and if $I = [\alpha, \beta]$ is an interval with midpoint γ, then

$$\int_\alpha^\beta P(x)\,dx = \frac{\beta - \alpha}{6}\left\{P(\alpha) + 4P(\gamma) + P(\beta)\right\}.$$

Proof. We have

$$
\begin{aligned}
\int_\alpha^\beta P(x)\,dx &= \int_\alpha^\beta Ax^2 + Bx + C\,dx \\
&= \frac{A}{3}(\beta^3 - \alpha^3) + \frac{B}{2}(\beta^2 - \alpha^2) + C(\beta - \alpha) \\
&= \frac{\beta - \alpha}{6}\left\{2A(\beta^2 + \beta\alpha + \alpha^2) + 3I(\beta + \alpha) + 6C\right\}
\end{aligned}
$$

$$= \frac{\beta - \alpha}{6} \left\{ \left[A\alpha^2 + B\alpha + C \right] \right.$$
$$+ 4 \left[A \left(\frac{\alpha + \beta}{2} \right)^2 + B \left(\frac{\alpha + \beta}{2} \right) + C \right]$$
$$\left. + \left[A\beta^2 + B\beta + C \right] \right\}$$
$$= \frac{\beta - \alpha}{6} \left\{ P(\alpha) + 4P(\gamma) + P(\beta) \right\}.$$

This completes the proof.   □

We note in passing that, since the equation of a parabola

$$y = Ax^2 + Bx + C$$

has three coefficients, then the parabola is uniquely determined by any three points through which it passes.

Now, to derive Simpson's Rule, we take a given function which we wish to integrate over an interval $[a, b]$, and we select a partition with *an even number of intervals, of equal length* Δx. This is shown in Figure 2.10.

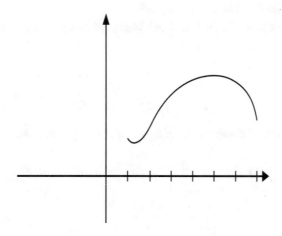

Figure 2.10: Applying Simpson's Rule.

We pair up the intervals (Figure 2.10), and on each pair of intervals, we approximate f by a parabola passing through the points on the graph of f corresponding to the three endpoints of the intervals (Figure 2.11).

If $k = 2\ell$ is the number of intervals, then a typical pair of intervals is

$$[x_{2j-2}, x_{2j-1}] \quad , \quad [x_{2j-1}, x_{2j}] \ , \quad j = 1, \ldots, \ell.$$

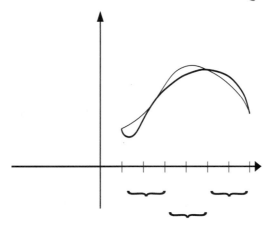

Figure 2.11: Pairing up the intervals.

The approximating parabola $P_j(x)$ passes through the points

$$(x_{2j-2}, f(x_{2j-2})) \ , \ (x_{2j-1}, f(x_{2j-1})) \ , \ (x_{2j}, f(x_{2j})) \ .$$

We apply the lemma, with $\alpha = x_{2j-2}, \gamma = x_{2j-1}, \beta = x_{2j}$, to obtain that the integral of the parabola over the interval $[x_{2j-2}, x_{2j}]$ is

$$\frac{x_{2j} - x_{2j-2}}{6}\{P_j(x_{2j-2}) + 4P_j(x_{2j-1}) + P_j(x_{2j})\}$$
$$= \frac{2\Delta x}{6}\{f(x_{2j-2}) + 4f(x_{2j-1}) + f(x_{2j})\} \ .$$

Finally, we add up these integrals, from $j = 1$ to $j = \ell$, to obtain the approximation

$$\int_a^b f(x)\,dx \ \approx \ \frac{\Delta x}{3}\{f(x_0) + 4f(x_1) + 2f(x_2) + 4f(x_3) + 2f(x_4) + \ldots$$
$$+ 2f(x_{k-2}) + 4f(x_{k-1}) + f(x_k)\}.$$

We summarize Simpson's method:

Theorem 2.20 (Simpson's Rule)   *Let f be a continuous function whose domain contains the interval $[a, b]$. Let $\{x_1, x_2, \ldots, x_k\}$ be a uniform partition of the interval, with each subinterval having length Δx and with $k = 2\ell$ being even. The integral*

$$\int_a^b f(x)\,dx$$

is approximated by the expression

$$\frac{\Delta x}{3} \left\{ f(x_0) + 4f(x_1) + 2f(x_2) + 4f(x_3) + \dots \ 2f(x_{k-2}) + 4f(x_{k-1}) + f(x_k) \right\}.$$

It can be determined that, if $|f^{(iv)}(x)| \leq M$, then the error in applying Simpson's method is not greater than

$$\frac{M(b-a)^5}{180k^4}.$$

Example 2.21   Redo Example 2.13 with the same partition $\{0, 1/4, 1/2, 3/4, 1\}$, using Simpson's Rule.

It is straightforward to calculate, since $f(x) = 1/(1 + x^2)$, that

$$f^{(iv)}(x) = \frac{120x^4 - 240x^2 + 24}{(1 + x^2)^5}.$$

It can be determined that $f^{(iv)}$ takes its maximum value on the interval $[0, 1]$ at $x = 0$. Therefore,

$$|f^{(iv)}| \leq 24.$$

Hence, with $k = 4$, Simpson's Rule should be accurate to within

$$\frac{24 \cdot 1^5}{180 \cdot 4^4} \approx 0.000520833.$$

This is slightly worse than three decimal places of accuracy.

Now we estimate the integral using Simpson's Rule:

$$\int_0^1 \frac{1}{1+x^2} \, dx \approx \frac{0.25}{3} \left\{ 1 + 4 \cdot \frac{16}{17} + 2 \cdot \frac{4}{5} + 4 \cdot \frac{16}{25} + \frac{1}{2} \right\} = \frac{0.25}{3} \cdot \frac{8011}{850} \approx 0.78539216.$$

Noting from Example 2.13 that the true value of the integral is

$$0.78539816,$$

we see that we have actually achieved four decimal place accuracy.

Remark 2.22   We know from calculus that the *exact value* of the integral

$$\int_0^1 \frac{1}{1+x^2} \, dx$$

is $\pi/4$. In Example 2.21, using only arithmetic, we approximated the value of the integral to four decimal place accuracy by the quantity

$$0.78539216.$$

Therefore, Example 2.21 gives the following approximation to the quantity π :

$$\pi = 4 \cdot \frac{\pi}{4} \approx 4 \cdot 0.78539216 = 3.14156864.$$

To four decimal place accuracy, this gives the value of π to be 3.1416. This is one of the many ways that the value of π can be calculated by hand.

Example 2.23   Using Simpson's Rule, how fine a partition is required to estimate the integral

$$\int_0^1 e^{-x^2}\,dx$$

to four decimal places of accuracy?

After a calculation, we find that

$$\frac{d^4}{dx^4}\left\{e^{-x^2}\right\} = e^{-x^2}(16x^4 - 48x^2 + 12).$$

Now the maximum absolute value of the function $|e^{-t} \cdot (16t^2 - 48t + 12)|$ on the interval $[0, 1]$ is easily seen to occur at the origin. Thus, the maximum absolute value of the fourth derivative of the integrand is $M = 12$. According to Theorem 2.20, we need to choose k such that

$$\frac{12(1-0)^5}{180k^4} = \frac{M(b-a)^5}{180k^4} < 5 \cdot 10^{-5}.$$

This simplifies to

$$\frac{12 \cdot 1 \cdot 10^5}{180 \cdot 5} < k^4.$$

A quick calculation shows that $k = 8$ satisfies this inequality (recall that k must be even in Simpson's rule).

In conclusion, we have found that with relatively few calculations (compare Examples 2.15 and 2.18), we may achieve excellent accuracy.

2.6   INTEGRATION BY PARTS

The formula

$$\frac{d}{dx}(u \cdot v) = \frac{du}{dx} \cdot v + u \cdot \frac{dv}{dx}$$

is familiar from calculus. We may rearrange it to read

$$u \cdot \frac{dv}{dx} = \frac{d}{dx}(u \cdot v) - \frac{du}{dx} \cdot v \,.$$

Now apply the integral to both sides, using the Fundamental Theorem of Calculus on the left, to obtain

$$\int u \cdot \frac{dv}{dx}\, dx = \int \frac{d}{dx}(u \cdot v)\, dx - \frac{du}{dx} \cdot v\, dx$$

or, using the formalism of calculus,

$$\int u\, dv = uv - \int v\, du\,.$$

This is the integration by parts formula of Newton and Leibniz. It is a powerful tool in mathematical analysis.

There is a discrete analogue of integration by parts that is perhaps worth mentioning now. It is called *summation by parts*.

Proposition 2.24   (Summation by Parts)   *Let $\{a_j\}_{j=0}^{\infty}$ and $\{b_j\}_{j=0}^{\infty}$ be two sequences of real numbers. For $N = 0, 1, 2, \ldots$ set*

$$A_N = \sum_{j=0}^{N} a_j$$

(we adopt the convention that $A_{-1} = 0$.) Then, for any $0 \leq m \leq n < \infty$, it holds that

$$\sum_{j=m}^{n} a_j \cdot b_j = \left[A_n \cdot b_n - A_{m-1} \cdot b_m \right]$$

$$+ \sum_{j=m}^{n-1} A_j \cdot (b_j - b_{j+1})\,.$$

Proof. We write

$$\begin{aligned}
\sum_{j=m}^{n} a_j \cdot b_j &= \sum_{j=m}^{n} \left(A_j - A_{j-1}\right) \cdot b_j \\
&= \sum_{j=m}^{n} A_j \cdot b_j - \sum_{j=m}^{n} A_{j-1} \cdot b_j \\
&= \sum_{j=m}^{n} A_j \cdot b_j - \sum_{j=m-1}^{n-1} A_j \cdot b_{j+1} \\
&= \sum_{j=m}^{n-1} A_j \cdot (b_j - b_{j+1}) + A_n \cdot b_n - A_{m-1} \cdot b_m \,.
\end{aligned}$$

This is what we wished to prove. □

Now we apply summation by parts to prove a convergence test due to Niels Henrik Abel (1802-1829).

Theorem 2.25 (Abel's Convergence Test) *Consider the series*

$$\sum_{j=0}^{\infty} a_j \cdot b_j \,.$$

Suppose that

1. The partial sums $A_N = \sum_{j=0}^{N} a_j$ form a bounded sequence;

2. The b_j are nonnegative and $b_0 \geq b_1 \geq b_2 \geq \cdots$;

3. $\lim_{j \to \infty} b_j = 0$.

Then the original series

$$\sum_{j=0}^{\infty} a_j \cdot b_j$$

converges.

Proof. Suppose that the partial sums A_N are bounded in absolute value by a number K. Pick $\epsilon > 0$, and choose an integer N so large that $b_N < \epsilon/(2K)$. For $N \leq m \leq n < \infty$, we use the partial summation formula to write

$$\left| \sum_{j=m}^{n} a_j \cdot b_j \right| = \left| A_n \cdot b_n - A_{m-1} \cdot b_m + \sum_{j=m}^{n-1} A_j \cdot (b_j - b_{j+1}) \right|$$

$$\leq K \cdot |b_n| + K \cdot |b_m| + K \cdot \sum_{j=m}^{n-1} |b_j - b_{j+1}| .$$

Now we take advantage of the facts that $b_j \geq 0$ for all j and that $b_j \geq b_{j+1}$ for all j to estimate the last expression by

$$K \cdot \left[b_n + b_m + \sum_{j=m}^{n-1} \left(b_j - b_{j+1} \right) \right] .$$

[Notice that the expressions $b_j - b_{j+1}$, b_m, and b_n are all positive.] Now the sum collapses and the last line is estimated by

$$K \cdot [b_n + b_m - b_n + b_m] = 2 \cdot K \cdot b_m .$$

By our choice of N, the right side is smaller than ϵ. Thus, our series satisfies the Cauchy criterion and therefore converges.   □

Example 2.26   Let us apply Abel's test to see that an alternating series converges.
Consider the series

$$\sum_{j=1}^{\infty} (-1)^j b_j ,$$

where the b_j are nonnegative real numbers that decrease monotonically to zero. We see immediately, with $a_j = (-1)^j$ and $b_j = b_j$, that Abel's test applies and the series converges.

EXERCISES

1. Let f, g be Riemann integrable functions on the interval $[a, b]$. Prove that

$$h(x) = \max\{f(x), g(x)\}$$

is also integrable on $[a, b]$. Prove likewise for the minimum.

2. Define a concept of *upper Riemann integral* (where the ξ_j that you select in each $[x_{j-1}, x_j]$ is the location of the maximum for the function f) and also define a concept of *lower Riemann integral* (where the ξ_j that you select in each $[x_{j-1}, x_j]$ is the location of the minimum for the function f). Calculate the upper and lower Riemann sums for the Dirichlet function

$$f(x) = \begin{cases} 1 & \text{if} \quad x \text{ is rational} \\ 0 & \text{if} \quad x \text{ is irrational} . \end{cases}$$

3. Let $Q = \{q_1, q_2, \ldots, q_k\}$ be a collection of finitely many rational numbers. Explain why the function

$$f_Q(x) = \begin{cases} 1 & \text{if} \quad x \in Q \\ 0 & \text{if} \quad x \notin Q \end{cases}$$

is integrable, yet the function

$$g(x) = \begin{cases} 1 & \text{if} \quad x \in Q \\ 0 & \text{if} \quad x \notin Q \end{cases}$$

is not integrable.

4. Let

$$h(x) = \begin{cases} x \cdot \sin(1/x) & \text{if} \quad x \neq 0 \\ 0 & \text{if} \quad x = 0 \end{cases}$$

on the interval $[-1, 1]$. Is h Riemann integrable? Why or why not?

5. The integral

$$\int_0^\infty \frac{\sin x}{x} \, dx$$

is what is called an *improper Riemann integral*. But we may make sense of it by evaluating

$$\lim_{N \to +\infty} \int_{1/N}^N \frac{\sin x}{x} \, dx \, .$$

Do so.

6. Provide the details of the assertion that, if f is Riemann integrable on the interval $[a, b]$, then for any $\epsilon > 0$ there is a $\delta > 0$ such that if $\mathcal{P}$ is a partition of mesh less than δ then

$$\sum_j \left(\sup_{I_j} f - \inf_{I_j} f \right) \Delta_j < \epsilon \, .$$

[**Hint:** Follow this scheme: Given $\epsilon > 0$, choose $\delta > 0$ as in the definition of the integral. Fix a partition $\mathcal{P}$ with mesh smaller than δ. Let $K + 1$ be the number of points in $\mathcal{P}$. Choose points $t_j \in I_j$ so that $|f(t_j) - \sup_{I_j} f| < \epsilon/(2(K+1))$; also choose points $t'_j \in I_j$ so that $|f(t'_j) - \inf_{I_j} f| < \epsilon/(2(K+1))$. By applying the definition of the integral to this choice of t_j and t'_j, we find that

$$\sum_j \left(\sup_{I_j} f - \inf_{I_j} f \right) \Delta_j < 2\epsilon \, .$$

The result follows.]

7. Let f be a bounded function on an unbounded interval of the form $[A, \infty)$. We say that f is integrable on $[A, \infty)$ if f is integrable on every compact subinterval of $[A, \infty)$ and

$$\lim_{B \to +\infty} \int_A^B f(x)\, dx$$

exists and is finite.

Assume that f is Riemann integrable on $[1, N]$ for every $N > 1$ and that f is monotone decreasing. Show that f is Riemann integrable on $[1, \infty)$ if and only if $\sum_{j=1}^{\infty} f(j)$ is finite.

Suppose that g is nonnegative and integrable on $[1, \infty)$. If $0 \le |f(x)| \le g(x)$ for $x \in [1, \infty)$, then prove that f is integrable on $[1, \infty)$.

8. Let f be a function on an interval of the form $(a, b]$ such that f is integrable on compact subintervals of $(a, b]$. If

$$\lim_{\epsilon \to 0^+} \int_{a+\epsilon}^b f(x)\, dx$$

exists and is finite then we say that f is integrable on $(a, b]$. Prove that, if we restrict attention to bounded f, then in fact this definition gives rise to no new integrable functions. However, there are unbounded functions that can now be integrated. Give an example.

Give an example of a function g that is integrable by the definition in the preceding paragraph but is such that $|g|$ is not integrable.

9. Give an example to show that the composition of Riemann integrable functions need not be Riemann integrable.

10. Suppose that f is a continuous, nonnegative function on the interval $[0, 1]$. Let M be the supremum of f on the interval. Prove that

$$\lim_{n \to \infty} \left[\int_0^1 f(t)^n\, dt \right]^{1/n} = M.$$

11. Let f be a continuous function on the interval $[0, 1]$ that only takes nonnegative values there. Prove that

$$\left[\int_0^1 f(t)\, dt \right]^2 \le \int_0^1 f(t)^2\, dt.$$

12. To what extent is the following statement true? If f is Riemann integrable on $[a, b]$, then $1/f$ is Riemann integrable on $[a, b]$.

13. Explain how the summation by parts formula may be derived from the integration by parts formula.

14. Explain how the integration by parts formula may be derived from the summation by parts process.

15. Give an example of a function f such that f^2 is Riemann integrable, but f is not. What additional hypothesis on f would make the implication true?

16. Let f be a continuously differentiable function on the interval $[0, 2\pi]$. Further assume that $f(0) = f(2\pi)$ and $f'(0) = f'(2\pi)$. For $n \in \mathcal{N}$, define

$$\widehat{f}(n) = \frac{1}{2\pi} \int_0^{2\pi} f(x) \sin nx \, dx .$$

Prove that

$$\sum_{n=1}^{\infty} |\widehat{f}(n)|^2$$

converges. [**Hint:** Use integration by parts to obtain a favorable estimate on $|\widehat{f}(n)|$.]

17. Prove that

$$\lim_{\eta \to 0^+} \int_{\eta}^{1/\eta} \frac{\cos(2r) - \cos r}{r} dr$$

exists.

18. If f is Riemann integrable on the interval $[a, b]$ and if $\mu : [\alpha, \beta] \to [a, b]$ is continuous, then prove that $f \circ \mu$ is Riemann integrable on $[\alpha, \beta]$.

19. Is the product of Riemann integrable functions Riemann integrable? Why or why not? If f^3 is Riemann integrable, does it then follow that f is Riemann integrable?

20. Prove that the uniform limit of Riemann integrable functions is Riemann integrable.

CHAPTER 3

The Lebesgue Integral

3.1 ELEMENTARY MEASURE THEORY

The Lebesgue integral is based on a theory of measure. Measure theory was created with the idea that we would like to have a way to measure the length of any subset $E \subseteq \mathcal{R}$. It turns out that it is logically impossible to do so.

We should say in advance what sorts of properties we would like a measure to have. Here are some of them:

(i) The measure of any set should be a nonnegative number.

(ii) Any interval $[a, b]$ should have length or measure $b - a$.

(iii) If a set E has measure β, then any translate $E_a = \{x + a : x \in E\}$ of E by a units should also have measure β.

(iv) If A_j are pairwise disjoint sets and $A = \cup_j A_j$, then the measure of A should be the sum of the measures of the A_j.

The next example illustrates the difficulty of attempting to measure all sets consistent with the four desiderata enunciated above.

Example 3.1   There is a subset $E \subseteq [0, 1]$ to which it is impossible to assign a measure or length. To see this, let us define a relation on $[0, 1]$ by $x \sim y$ if $x - y$ is rational. This is, of course, an equivalence relation, and we may consider the equivalence classes. There are uncountably many of these. Let E be a set consisting of one element from each equivalence class (the Axiom of Choice is needed in order to perform this last step).

We assert that it is impossible to assign a measure to E. For we may consider the sets, for $q \in [0, 1]$ rational, given by

$$E_q = \{x + q : x \in E\},$$

where the addition is performed modulo 1. These sets are pairwise disjoint. And their union is the interval $[0, 1]$. According to our four purported rules for a measure (particularly rule **(iii)**), the sets E_q each have the same measure as E. If E has zero measure, then each E_q must have zero measure for each q, and therefore $[0, 1]$, being the countable union of sets of zero measure, must have measure zero. That is false because the measure of $[0, 1]$ must be 1. If instead E has positive measure $\alpha > 0$

then each E_q has measure α and hence $[0, 1]$, being the countable union of sets of measure α, must have infinite measure. That too is impossible.

One might infer from this example that it is unwise to allow countable additivity of the measure. But the following example in dimension three shows that is not the difficulty.

Example 3.2 (Banach-Tarski) Let A be the unit ball in $\mathcal{R}^3$, and let B be the Empire State building. There are finitely many pairwise disjoint sets, $E_1, E_2, \ldots, E_k$, and finitely many pairwise disjoint sets, $F_1, F_2, \ldots, F_k$, such that each E_j is geometrically congruent to F_j, $j = 1, \ldots, k$, and furthermore

$$A = E_1 \cup \cdots \cup E_k \qquad \text{and} \qquad B = F_1 \cup \cdots F_k.$$

This is really counterintuitive. Surely, the sets E_j and F_j are pathological. It requires the Axiom of Choice to construct them.

We cannot provide all the details of this Banach-Tarski Paradox, but instead refer the reader to the splendid exposition in [JEC].

Thus, we must restrict attention to a class of sets called "measurable sets." Likewise, we shall later restrict our focus to a class of functions called "measurable functions."

3.2   MEASURABLE SETS

Let $S \subseteq \mathcal{R}$ be any set. An *open covering* of S is a collection $\mathcal{U} = \{U_j\}_{j \in \mathcal{N}}$ of open intervals with the property that $\cup_j U_j \supseteq S$. Of course, each interval $U_j \equiv (a_j, b_j)$ has a natural length $b_j - a_j$. We may say that the *length* $\ell(\mathcal{U})$ of the covering is $\sum_j (b_j - a_j)$. We set

$$m^*(S) = \inf_{\mathcal{U} \text{ a covering of } S} \ell(\mathcal{U}).$$

This defines an *outer measure*. This outer measure will not satisfy the four axioms that we set out above. But it gives a rough means of measuring the length of a set.

Example 3.3 Let E be the rational numbers $\mathcal{Q}$ in the reals. Let us determine the outer measure of E. Of course, the rational numbers are countable, so let us write $\mathcal{Q} = \{q_1, q_2, \ldots\}$. Let $\epsilon > 0$. Now define

$$
\begin{aligned}
I_1 &= (q_1 - \epsilon/4, q_1 + \epsilon/4) \\
I_2 &= (q_2 - \epsilon/8, q_2 + \epsilon/8) \\
I_3 &= (q_3 - \epsilon/16, q_3 + \epsilon/16) \\
&\cdots \\
I_k &= (q_k - \epsilon/2^{k+1}, q_k + \epsilon/2^{k+1}).
\end{aligned}
$$

Then

$$E \subseteq \bigcup_{j=1}^{\infty} I_j \equiv \mathcal{U} \,,$$

and $\mathcal{U}$ is open. Note also that $\ell(\mathcal{U}) \leq \epsilon$.

Since this is true for ever $\epsilon > 0$, we conclude that $m(E) = 0$.

As previously indicated, we can only have a workable measure theory if we restrict attention to a class of reasonable sets. This class should be closed under countable intersection and countable union. In fact, we formalize this idea as follows:

Definition 3.4   A σ-algebra on $\mathcal{R}$ is a collection of subsets of $\mathcal{R}$ that is closed under countable union and complementation (and therefore also under countable intersection).

We define the class of measurable sets as follows.

Definition 3.5   Say that a set $A \subseteq \mathcal{R}$ is *measurable* if, for any set $E \subseteq \mathcal{R}$,

$$m^*(E) = m^*(E \cap A) + m^*(E \cap {}^c A) \,. \tag{$*$}$$

We write $A \in \mathcal{M}$.

We now have the following central result of Carathéodory.

Theorem 3.6   *Let m be an outer measure as defined above. The collection of measurable sets, defined according to equation ($*$), forms a σ-algebra.*

Proof. The definition of measurable set is symmetric in A and ${}^c A$. Hence, the measurable sets are clearly closed under complementation.

Now let $A, B \in \mathcal{M}$ and let E be an arbitrary subset of $\mathcal{R}$. By subadditivity of m, we have

$$
\begin{aligned}
m^*(E) &= m^*(E \cap A) + m^*(E \cap {}^c A) \\
&= m^*(E \cap A \cap B) + m^*(E \cap A \cap {}^c B) \\
&\quad + m^*(E \cap {}^c A \cap B) + m^*(E \cap {}^c A \cap {}^c B) \\
&\geq m^*(E \cap (A \cup B)) + m^*(E \cap {}^c(A \cup B)) \,.
\end{aligned}
$$

We see then that $A \cup B \in \mathcal{M}$. Furthermore, if $A, B \in \mathcal{M}$ and $A \cap B = \emptyset$, then

$$
\begin{aligned}
m^*(A \cup B) &= m^*((A \cup B) \cap A) + m^*((A \cup B) \cap {}^c A) \\
&= m^*(A) + m^*(B) \,.
\end{aligned}
$$

So m is finitely additive.

To show that $\mathcal{M}$ is a σ-algebra, it suffices now to show that $\mathcal{M}$ is closed under countable pairwise disjoint unions. So let $\{A_j\}$ be a countable sequence of pairwise disjoint sets in $\mathcal{M}$. Set $B_n = \cup_{j=1}^{n} A_j$ and $B = \cup_{j=1}^{\infty} A_j$. For any $E \subseteq \mathcal{R}$, we see then that

$$
\begin{aligned}
m^*(E \cap B_n) &= m^*(E \cap B_n \cap A_n) + m^*(E \cap B_n \cap {}^c A_n) \\
&= m^*(E \cap A_n) + m^*(E \cap B_{n-1}) \,.
\end{aligned}
$$

Hence, a simple induction shows that $m^*(E \cap B_n) = \sum_{j=1}^{n} m^*(E \cap A_j)$. Therefore,

$$
\begin{aligned}
m^*(E) &= m^*(E \cap B_n) + m^*(E \cap {}^c B_n) \\
&\geq \left[\sum_{j=1}^{n} m^*(E \cap A_j) \right] + m^*(E \cap {}^c B) \,.
\end{aligned}
$$

Letting $n \to \infty$ gives

$$
\begin{aligned}
m^*(E) &\geq \left[\sum_{j=1}^{\infty} m^*(E \cap A_j) \right] + m^*(E \cap {}^c B) \\
&\geq m \left(\bigcup_{j=1}^{\infty} (E \cap A_j) \right) + m^*(E \cap {}^c B) \\
&= m^*(E \cap B) + m^*(E \cap {}^c B) \\
&\geq m^*(E) \,.
\end{aligned}
$$

The first and last terms of this string of inequalities are the same. We therefore conclude that these are, in fact, all equalities. Thus, $B \in \mathcal{M}$. Also, taking $E = B$, we see that

$$
m^*(B) = \sum_{j=1}^{\infty} m^*(A_j) \,.
$$

Hence, m is countably additive on $\mathcal{M}$.   □

Thus, we have what we want: A collection of sets that can be measured, and that are closed under reasonable operations. It is also important to know that this collection contains certain basic sets. We have

Proposition 3.7   *Every open set is measurable.*

Remark 3.8   It follows from this proposition and the preceding theorem that every closed set is measurable. Indeed, any set that can be obtained from the open sets by way of complementation and countable union (and hence also by way of countable intersection) is measurable. The collection of such subsets is called the *Borel sets*, and they are a central part of measure theory.

Proof of the Proposition. We need not consider an arbitrary open set A. In fact, it is enough to consider A an open halfline. Let E be an arbitrary set in $\mathcal{R}$.

Let $\mathcal{U} = \{U_j\}$ be any countable open covering of E, chosen so that

$$|\ell(\mathcal{U}) - m^*(E)| < \epsilon.$$

Then $\{U_j \cap A\} \equiv \{V_j\} = \mathcal{V}$ is an open covering of $E \cap A$ and surely $|\ell(\mathcal{V}) - m^*(E \cap A)| < \epsilon$. Of course, the complement ^cA of A is a closed halfline. Then $\{U_j \cap {}^cA\} \equiv \{W_j\} \equiv \mathcal{W}$, together with an interval of length less than $\epsilon/2$ that covers the endpoint of ^cA gives a cover of $E \cap {}^cA$ so that

$$|\ell(\mathcal{W}) - m^*(E \cap {}^cA)| < \epsilon.$$

It follows that

$$m^*(E) \geq m^*(E \cap A) + m^*(E \cap {}^cA) - 2\epsilon.$$

Since $\epsilon > 0$ was arbitrary, we conclude that

$$m^*(E) \geq m^*(E \cap A) + m^*(E \cap {}^cA).$$

The inequality in the opposite direction comes for free from subadditivity, so we conclude that

$$m^*(E) = m^*(E \cap A) + m^*(E \cap {}^cA).$$

Hence, A is measurable. $\qquad\square$

It is common in measure theory to say that a certain phenomenon is true "almost everywhere" if it is true except for a set of points having measure zero. For example, in what follows, we shall have occasion to say that $f(x) = g(x)$ almost everywhere. This means that $\{x \in \mathcal{R} : f(x) \neq g(x)\}$ has measure zero. We also write $f(x) = g(x)$ a.e.

In what follows, when we are discussing the measure of a measurable set E, we shall write $m(E)$ (in other words, we shall omit the $*$).

3.3 THE LEBESGUE INTEGRAL

The Riemann integral is built on the idea of dividing up the domain of the function, or integrand. The Lebesgue integral is built on the idea of dividing up the range. We begin with some definitions and observations.

In what follows, we shall follow the convention that $0 \cdot \infty = 0$. This is reasonable because usually ∞ arises as the limit of an increasing sequence of finite real numbers (such as partial sums of a series), and certainly $0 \cdot N = 0$ for any real N.

Definition 3.9 Let S be a measurable set. Then the function

$$\chi_S(x) = \begin{cases} 1 & \text{if} & x \in S \\ 0 & \text{if} & x \notin S \end{cases}$$

is called the *characteristic function* of S.

Definition 3.10   Let $E_1, E_2, ..., E_k$ be measurable sets of finite measure. Let $\alpha_1, \alpha_2, ..., \alpha_k$ be real constants. Then the function

$$f(x) = \sum_{j=1}^{k} \alpha_j \chi_{E_j}(x)$$

is called a *simple function.*

Definition 3.11   Let

$$f(x) = \sum_{j=1}^{k} \alpha_j \chi_{E_j}(x)$$

be a simple function. Then the Lebesgue integral of f is given by

$$\int_{\mathcal{R}} f(x)\,dx = \sum_{j=1}^{k} \alpha_j m(E_j).$$

Definition 3.12   A function $f : \mathcal{R} \to \mathcal{R}$ is called *measurable* if, whenever $U \subseteq \mathcal{R}$ is an open set, then $f^{-1}(U)$ is measurable.

Remark 3.13   It is worth comparing this last definition to the definition of continuous function from topology. In the subject of topology, we say that a function $f : X \to Y$ is continuous if the inverse image of any open set is open. The definition of measurable function is a variant of that idea.

In our consideration of measurable functions, it is useful to allow functions to take the values $\pm\infty$. The Borel sets in the extended real line $\overline{\mathcal{R}} \equiv \mathcal{R} \cup \{-\infty, +\infty\}$ are generated by the halflines $[-\infty, a)$ and $(b, \infty]$ using countable union, countable intersection, and complementation. We call $\overline{\mathcal{R}}$ the *extended real numbers*. We say that a function taking values in $\overline{\mathcal{R}}$ is measurable if the inverse image of any of these halflines is a measurable set in the usual sense.

Example 3.14   Let S be a measurable set, and let $f(x) = \chi_S(x)$. Then f is a measurable function.
    To see this, notice that

- If $\lambda < 0$, then

$$f^{-1}(\{x \in \mathcal{R} : f(x) < \lambda\})$$

is the empty set.

- If $0 \leq \lambda < 1$, then

$$f^{-1}(\{x \in \mathcal{R} : f(x) < \lambda\})$$

is the set ${}^c S$.

- If $1 \leq \lambda < \infty$, then

$$f^{-1}(\{x \in \mathcal{R} : f(x) < \lambda\})$$

is all of $\mathcal{R}$.

- If $\lambda < 0$, then

$$f^{-1}(\{x \in \mathcal{R} : f(x) > \lambda\})$$

is all of $\mathcal{R}$.

- If $0 \leq \lambda < 1$, then

$$f^{-1}(\{x \in \mathcal{R} : f(x) > \lambda\})$$

is S.

- If $\lambda \geq 1$, then

$$f^{-1}(\{x \in \mathcal{R} : f(x) > \lambda\})$$

is the empty set.

We note that each of these inverse images is measurable. Any open interval is the intersection of two open halflines, so its inverse image will be measurable. And any open set is the countable union of open intervals, so its inverse image will be measurable.

Example 3.15   Any continuous function is measurable because any open set is measurable.

Example 3.16   Any simple function is measurable. Now a simple function only takes finitely many values. So we can break up the range just as we did in Example 3.14 to see that the inverse image of any open halfline is measurable. From this, it follows that the inverse image of any open set is measurable.

As indicated earlier, in analysis we are interested in limiting processes. So we naturally want to know that the class of measurable functions is closed under limits. We have:

Proposition 3.17   *Let f_j be a sequence of measurable functions. Then the functions*

$$g(x) \equiv \sup_j f_j(x) \qquad\qquad h(x) \equiv \inf_j f_j(x)$$

$$m(x) = \limsup_{j \to \infty} f_j(x) \qquad\qquad n(x) \equiv \liminf_{j \to \infty} f_j(x)$$

are all measurable.

Corollary 3.18   *If f_j are measurable functions and $f_j(x) \to f(x)$ almost everywhere, then f is measurable.*

Proof of the Proposition. Certainly

$$g^{-1}((a, \infty)) = \bigcup_{j=1}^{\infty} f_j^{-1}((a, \infty))$$

and

$$h^{-1}((-\infty, a)) = \bigcap_{j=1}^{\infty} f_j^{-1}((-\infty, a)).$$

So these sets are measurable, and it follows that g, h are measurable.

Furthermore, if we set

$$p_k(x) = \sup_{j > k} f_j(x)$$

then p_k is measurable for each k, hence $m \equiv \inf_{k \geq 1} p_k$ is measurable. The function n is treated in the same way. □

Now, of course, simple functions are a bit too rudimentary to be of any practical use. So we must find a way to extend our definition of the integral from simple functions to arbitrary measurable functions. We do so as follows:

Proposition 3.19   *Let f be a nonnegative, measurable function. Then there is a sequence*

$$s_1 \leq s_2 \leq s_3 \leq \cdots$$

of simple functions such that

$$\lim_{j \to +\infty} s_j(x) = f(x)$$

for every $x \in \mathcal{R}$. Moreover, $s_j \to f$ uniformly on any set on which f is bounded.

Proof. If n is a natural number and $0 \leq k \leq 2^{2n} - 1$, then let

$$E_n^k = f^{-1}\big((k \cdot 2^{-n}, (k+1) \cdot 2^{-n}]\big) \qquad \text{and} \qquad F_n = f^{-1}\big((2^n, \infty]\big).$$

Set

$$s_n = \left[\sum_{j=0}^{2^{2n}-1} k \cdot 2^{-n} \chi_{E_n^k}\right] + 2^n \chi_{F_n}.$$

It is easy to see that $s_1 \leq s_2 \leq \cdots$ and also $0 \leq f - s_n \leq 2^{-n}$ on the set where $f \leq 2^n$. The result follows immediately. $\square$

Definition 3.20 Let f be a nonnegative, measurable function. Define

$$\int_{\mathcal{R}} f(x)\, dx = \sup \int_{\mathcal{R}} s(x)\, dx,$$

where the supremum is taken over the collection of all simple functions s such that $0 \leq s(x) \leq f(x)$ for all x.

It is clear that the proposition preceding this last definition gives the definition substance. For any nonnegative, measurable function can be well-approximated by simple functions.

Now if f is *any* measurable, real-valued function, then we set

$$f^+(x) = f(x) \cdot \chi_{\{x: f(x) > 0\}},$$

$$f^-(x) = -f(x) \cdot \chi_{\{x: f(x) < 0\}}.$$

Then we have:

Definition 3.21 If f is a measurable function, then we set

$$I^+ = \int_{\mathcal{R}} f^+(x)\, dx$$

and

$$I^- = \int_{\mathcal{R}} f^-(x)\, dx.$$

Finally, we define

$$\int_{\mathcal{R}} f(x)\, dx = I^+ - I^-$$

provided that both I^+ and I^- are finite.

In case $I^+ = \infty$ and I^- is finite, then we say that

$$\int_{\mathcal{R}} f(x)\,dx = +\infty.$$

In case $I^- = \infty$ and I_+ is finite, then we say that

$$\int_{\mathcal{R}} f(x)\,dx = -\infty.$$

In case *both* I^+ and I^- are infinite, then the integral does not exist.

In case $\int_{\mathcal{R}} f(x)\,dx$ exists and is finite, then we say that f is *integrable*. It is common to write $f \in L^1$ (which is a special case of the notation $f \in L^p$, to be discussed later).

Example 3.22   Define

$$f(x) = \begin{cases} \frac{1}{x^2} & \text{if} & x > 0 \\ -\frac{1}{x^2} & \text{if} & x < 0 \\ 0 & \text{if} & x = 0. \end{cases}$$

Then

$$f^+(x) = \frac{1}{x^2} \cdot \chi_{\{x:x>0\}}.$$

It is easy to see that the integral of f^+ is $+\infty$. Likewise, the integral of f^- is $+\infty$. Therefore, the Lebesgue integral of f does not exist.

Example 3.23   The *Dirichlet function* is

$$d(x) = \begin{cases} 1 & \text{if} & x \in [0,1], x \in \mathcal{Q} \\ 0 & \text{if} & x \notin [0,1] \text{ or } x \notin \mathcal{Q}. \end{cases}$$

So d is the characteristic function of the rational numbers in $[0,1]$. If we let $\mathcal{Q} = \{q_1, q_2, \dots\}$ be an enumeration of these rationals, then we may define

$$\mathcal{Q}_k = \{q_1, q_2, \dots, q_k\}$$

and let χ_k be the characteristic function of $\mathcal{Q}_k$. Then $\chi_k \nearrow \chi_{\mathcal{Q}}$.
    Then

$$\int_{\mathcal{R}} d(x)\,dx = \lim_{k \to \infty} \int_{\mathcal{R}} \chi_k(x)\,dx = 0.$$

So the Dirichlet function integrates to 0.

Example 3.24   Let

$$f(x) = \begin{cases} 1 & \text{if} & x \in [0,1], x \notin \mathcal{Q} \\ 0 & \text{if} & x \notin [0,1] \text{ or } x \in \mathcal{Q}. \end{cases}$$

Then

$$\int_{\mathcal{R}} f(x)\,dx = \int_{\mathcal{R}} \chi_{[0,1]} \cdot \left(1 - d(x)\right)dx,$$

where d is the Dirichlet function from the last example. This last line equals

$$1 - 0 = 1.$$

A basic fact about the integral, and one that we shall use frequently, is the following:

Proposition 3.25   *If f and g are integrable functions and $f(x) \le g(x)$ almost everywhere, then*

$$\int_{\mathcal{R}} f(x)\,dx \le \int_{\mathcal{R}} g(x)\,dx.$$

Proof. We may assume that f and g are both nonnegative. Let s_j be an increasing sequence of simple functions that converges to f. Then certainly $s_j \le g$ almost everywhere. Hence,

$$\int_{\mathcal{R}} s_j(x)\,dx \le \int_{\mathcal{R}} g(x)\,dx.$$

Passing to the limit now gives the result. □

3.4   THREE BIG THEOREMS ABOUT THE LEBESGUE INTEGRAL

As we have indicated, one of the seminal properties of the Lebesgue integral—that sets it apart from the Riemann integral—is the way that it behaves with respect to the limiting process. There are three key results about this behavior.

The first result is the most accessible. It is still quite useful.

Theorem 3.26   (The Lebesgue Monotone Convergence Theorem)   *Let $0 \le f_1 \le f_2 \le \cdots$ be measurable functions. Define $f(x) = \lim_{j \to \infty} f_j(x)$ (note that f could be equal to ∞ at some or all points). Then*

$$\lim_{j \to \infty} \int_{\mathcal{R}} f_j(x)\,dx = \int_{\mathcal{R}} f(x)\,dx.$$

Proof. Certainly

$$P_j = \int_{\mathcal{R}} f_j(x)\,dx$$

is an increasing sequence of nonnegative real numbers. So it has a limit ℓ (possibly equal to $+\infty$). Also it is clear that

$$\int_{\mathcal{R}} f_j(x)\,dx \le \int_{\mathcal{R}} f(x)\,dx$$

for all j. Thus,

$$\ell = \lim_{j \to \infty} \int_{\mathcal{R}} f_j(x)\,dx \le \int_{\mathcal{R}} f(x)\,dx.$$

To prove the reverse inequality, let $0 < \alpha < 1$. Also let s be a simple function with $0 \le s \le f$. Set

$$E_j = \{x \in \mathcal{R} : f_j(x) \ge \alpha s(x)\}.$$

Then E_j is an increasing sequence of measurable sets whose union is all of $\mathcal{R}$. Clearly,

$$\int_{\mathcal{R}} f_j(x)\,dx \ge \int_{E_j} f_j(x)\,dx \ge \alpha \cdot \int_{E_j} s(x)\,dx.$$

It is clear that $\lim_{j \to \infty} \int_{E_j} s(x)\,dx = \int_{\mathcal{R}} s(x)\,dx$. Therefore,

$$\lim_{j \to \infty} \int_{\mathcal{R}} f_j(x)\,dx \ge \alpha \cdot \int_{\mathcal{R}} s(x)\,dx.$$

Since this inequality is true for all $0 < \alpha < 1$, it is true in the limit as $\alpha \to 1$. Thus,

$$\lim_{j \to \infty} \int f_j(x)\,dx \ge \int_{\mathcal{R}} s(x)\,dx.$$

Taking the supremum over all simple functions s that lie below f, we find that

$$\lim_{j \to \infty} \int f_j(x)\,dx \ge \int_{\mathcal{R}} f(x)\,dx.$$

That is the desired inequality. $\qquad\square$

Of course, the finite additivity of the Lebesgue integral—that $\int (f + g)\,dx = \int f\,dx + \int g\,dx$—is a straightforward consequence of the definitions (see below). But what is more interesting is infinite additivity, and that is now easily derived from the Monotone Convergence Theorem:

Theorem 3.27   *Let f_j be nonnegative measurable functions. Set $f = \sum_j f_j$. Then*

$$\int_{\mathcal{R}} f(x)\,dx = \sum_j \int_{\mathcal{R}} f_j(x)\,dx.$$

Proof. We first treat the case of two functions. If f, g satisfy the hypotheses, then there are sequences of simple functions α_j and β_j which increase to f, g, respectively. Then $\alpha_j + \beta_j$ increases to $f + g$. The monotone convergence theorem then tells us that

$$\int_{\mathcal{R}} (f + g)\, dx = \lim_{j \to \infty} \int_{\mathcal{R}} (\alpha_j + \beta_j)\, dx = \lim_{j \to \infty} \int_{\mathcal{R}} \alpha_j\, dx + \lim_{j \to \infty} \int_{\mathcal{R}} \beta_j\, dx = \int_{\mathcal{R}} f\, dx + \int_{\mathcal{R}} g\, dx \,.$$

By induction, we may next establish that

$$\int_{\mathcal{R}} \left(\sum_{j=1}^{N} f_j(x) \right) dx = \sum_{j=1}^{N} \int_{\mathcal{R}} f_j(x)\, dx \,.$$

Now let $N \to \infty$ and apply the monotone convergence theorem once again. We conclude that

$$\int_{\mathcal{R}} \left(\sum_{j=1}^{\infty} f_j(x) \right) dx = \sum_{j=1}^{\infty} \int_{\mathcal{R}} f_j(x)\, dx \,.$$

That completes the proof. $\square$

Theorem 3.28 (Fatou's Lemma) *If f_j is any sequence of nonnegative, measurable functions, then*

$$\liminf_{j \to \infty} \int_{\mathcal{R}} f_j(x)\, dx \geq \int_{\mathcal{R}} \left(\liminf_{j \to \infty} f_j(x) \right) dx \,.$$

Proof. For each positive integer k, we have

$$\inf_{j \geq k} f_j(x) \leq f_\ell(x) \qquad \text{for } \ell \geq k \,.$$

Therefore,

$$\int_{\mathcal{R}} \inf_{j \geq k} f_j(x)\, dx \leq \int_{\mathcal{R}} f_\ell(x)\, dx \qquad \text{for } \ell \geq k \,.$$

It follows that

$$\int_{\mathcal{R}} \inf_{j \geq k} f_j(x)\, dx \leq \inf_{\ell \geq k} \int_{\mathcal{R}} f_\ell(x)\, dx \,.$$

Now let $k \to \infty$ and apply the monotone convergence theorem:

$$\int_{\mathcal{R}} \left(\liminf_{j \to \infty} f_j(x) \right) dx = \lim_{k \to \infty} \int_{\mathcal{R}} \left(\inf_{j \geq k} f_j(x) \right) dx \leq \liminf_{j \to \infty} \int_{\mathcal{R}} f_j(x)\, dx \,.$$

That is Fatou's lemma. $\square$

So far, all of our convergence theorems have been about nonnegative, measurable functions. Now we present the most powerful and flexible of the convergence theorems. It is about functions of any sign.

Theorem 3.29   *Let $\{f_j\}$ be a sequence of real-valued, measurable functions such that $f_j \to f$ pointwise. Suppose that there exists a nonnegative, integrable function g such that $|f_j(x)| \le g(x)$ for all x and all j. Then f is integrable and*

$$\lim_{j \to \infty} \int_{\mathcal{R}} f_j(x)\,dx = \int_{\mathcal{R}} f(x)\,dx\,.$$

Proof. Certainly f is measurable. We may apply Fatou's lemma to see that

$$\int_{\mathcal{R}} |f(x)|\,dx = \int_{\mathcal{R}} \liminf_{j \to \infty} |f_j(x)|\,dx \le \liminf_{j \to \infty} \int_{\mathcal{R}} |f_j(x)|\,dx \le \int_{\mathcal{R}} g(x)\,dx\,.$$

Hence, f is integrable.

As a result, $g + f_j \ge 0$ and $g - f_j \ge 0$ almost everywhere. Now Fatou's lemma tells us that

$$\int_{\mathcal{R}} g(x)\,dx + \int_{\mathcal{R}} f(x)\,dx \le \liminf_{j \to \infty} \int_{\mathcal{R}} \Big(g(x) + f_j(x)\Big)\,dx = \int_{\mathcal{R}} g(x)\,dx + \liminf_{j \to \infty} \int_{\mathcal{R}} f_j(x)\,dx$$

and

$$\int_{\mathcal{R}} g(x)\,dx - \int_{\mathcal{R}} f(x)\,dx \le \liminf_{j \to \infty} \int_{\mathcal{R}} \Big(g(x) - f_j(x)\Big)\,dx = \int_{\mathcal{R}} g(x)\,dx - \limsup_{j \to \infty} \int_{\mathcal{R}} f_j(x)\,dx\,.$$

Thus,

$$\liminf_{j \to \infty} \int_{\mathcal{R}} f_j(x)\,dx \ge \int_{\mathcal{R}} f(x)\,dx \ge \limsup_{j \to \infty} \int_{\mathcal{R}} f_j(x)\,dx\,,$$

and the result follows. $\qquad\square$

Example 3.30   As a simple application of dominated convergence, let us consider the Fourier transform

$$\widehat{f}(\xi) = \int_{\mathcal{R}} f(t)e^{it\xi}\,dt$$

of an integrable function f. We shall show that $\widehat{f}$ is continuous.

Fix a point $\xi_0 \in \mathcal{R}$. Notice that

$$|f(t)e^{it\xi}| \le |f(t)|$$

for every choice of ξ. So we apply dominated convergence with the role of g played by $|f|$. We see then that

$$\lim_{\xi \to \xi_0} \widehat{f}(\xi) = \lim_{\xi \to \xi_0} \int_{\mathcal{R}} f(t) e^{it\xi} \, dt = \int_{\mathcal{R}} \lim_{\xi \to \xi_0} \left(f(t) e^{it\xi} \right) dt = \int_{\mathcal{R}} f(t) e^{it\xi_0} \, dt = \widehat{f}(\xi_0) .$$

The result is proved.

3.5   THE LEBESGUE SPACES L^p

If $1 \le p < \infty$, then define $L^p(\mathcal{R})$ to be the collection of measurable functions f on $\mathcal{R}$ that satisfy

$$\int_{\mathcal{R}} |f(x)|^p \, dx < \infty .$$

In other words, an L^p function is p^{th} power integrable. We equip the space $L^p(\mathcal{R})$ with a norm as follows:

$$\|f\|_{L^p} = \int_{\mathcal{R}} |f(x)|^p \, dx^{1/p} .$$

In case $p = \infty$, the definition of L^p is a bit different. We set

$$\|f\|_{L^\infty} = \sup\left\{ \alpha > 0 : m(\{x : |f(x)| > \alpha\}) > 0 \right\} .$$

In other words, the L^∞ norm of f is the essential supremum of f.

The spaces $L^p(\mathcal{R})$ form a useful scale of spaces. We measure the utility of a linear operator in analysis by determining which of the L^p spaces it preserves.

It turns out that each of the L^p spaces is *complete*. That is to say, if $\{f_j\}$ is a Cauchy sequence in L^p, then the sequence has a limit in L^p. Also each of the L^p spaces is locally convex. We shall not prove either of these statements, as the proofs are technical and would take us far afield.

It is an interesting and useful fact that the Banach-space duals of the L^p spaces can be calculated. We have

Theorem 3.31   *Let $1 \le p < \infty$. Then the Banach-space dual of $L^p(\mathcal{R})$ is $L^{p'}(\mathcal{R})$, where $p' = p/(p-1)$.*

Again we shall omit the proof. It is also interesting to know that the dual of the continuous functions on a compact interval I is the space of finite Borel measures on I—see Section 3.6 on the Riesz representation theorem. There is no simple description of the dual of $L^\infty(\mathcal{R})$.

Two standard and useful tools in L^p theory are Hölder's inequality and Minkowski's inequality. We treat them now.

Theorem 3.32   (Hölder's Inequality)   *Let $1 \le p \le \infty$ and $q = p/(p-1)$. Let $f \in L^p$ and $g \in L^q$. Then*

$$\left| \int_{\mathcal{R}} f(x) \cdot g(x) \, dx \right| \le \int_{\mathcal{R}} |f(x)|^p \, dx^{1/p} \cdot \int_{\mathcal{R}} |g(x)|^q \, dx^{1/q} .$$

In other words,

$$\left| \int_{\mathcal{R}} f(x) \cdot g(x)\, dx \right| \leq \|f\|_{L^p} \cdot \|g\|_{L^q} .$$

Hölder's inequality was first proved in 1888 by L. J. Rogers. It was discovered independently in 1889 by Otto Hölder. In order to prove Hölder's inequality, we first need an elementary inequality of W. H. Young:

Lemma 3.33   *Let f be a strictly monotone increasing, continuous function on an interval $[0, c]$ with $f(0) = 0$. Let $a \in [0, c]$ and $b \in [0, f(c)]$. Then*

$$ab \leq \int_0^a f(x)\, dx + \int_0^b f^{-1}(x)\, dx .$$

Proof. The proof that we shall present is charming because it depends entirely on a picture. Examine Figure 3.1. We see that the area of the rectangle with side lengths a and b cannot be larger than the sum of the area under f, from 0 to a, and the area under f^{-1}, from 0 to b.   □

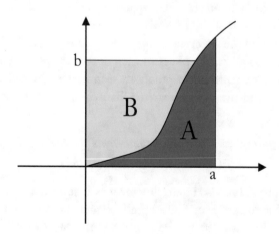

Figure 3.1: Young's inequality: $A = \int_0^a f(x)\, dx$ and $B = \int_0^b f^{-1}(x)\, dx$.

Proof of Hölder's Inequality. We can quickly dispose of a number of peripheral cases:

(a) If $\|f\|_{L^p} = 0$ or $\|g\|_{L^q} = 0$, then either $f = 0$ a.e. or $g = 0$ a.e. so that the left-hand side of Hölder's inequality is zero, so the inequality is trivially satisfied.

(b) If neither $\|f\|_{L^p} = 0$ nor $\|g\|_{L^1} = 0$ and instead either $\|f\|_{L^p} = \infty$ or $\|g\|_{L^q} = \infty$, then the right-hand side of Hölder's inequality is infinity. So the inequality is trivially satisfied.

(c) If $p = 1, q = \infty$ or $p = \infty, q = 1$, then the inequality is trivially confirmed as follows (in the first instance):

$$\left| \int_{\mathcal{R}} f(x) \cdot g(x) \, dx \right| \leq \int_{\mathcal{R}} |f(x)| \cdot |g(x)| \, dx$$

$$\leq \int_{\mathcal{R}} |f(x)| \cdot \|g\|_{L^\infty} \, dx$$

$$= \|g\|_{L^\infty} \cdot \int_{\mathcal{R}} |f(x)| \, dx$$

$$= \|g\|_{L^\infty} \cdot \|f\|_{L^1} \, .$$

Thus, we may consider $1 < p < \infty, 1 < q < \infty$, and we may assume that $\|f\|_{L^p}$ and $\|g\|_{L^q}$ are both finite. Replacing f by $f/\|f\|_{L^p}$ and g by $g/\|g\|_{L^q}$, we may suppose that $\|f\|_{L^p} = 1$ and $\|g\|_{L^q} = 1$. We now apply Young's inequality with $a = f(x)$ and $b = g(x)$. The result is

$$|f(x) \cdot g(x)| \leq \frac{|f(x)|^p}{p} + \frac{|g(x)|^q}{q} \, .$$

Now we integrate both sides in x. We calculate that

$$\int_{\mathcal{R}} |f(x) \cdot g(x)| \, dx \leq \frac{\int_{\mathcal{R}} |f(x)|^p \, dx}{p} + \frac{\int_{\mathcal{R}} |g(x)|^q \, dx}{q}$$

$$= \frac{1}{p} + \frac{1}{q}$$

$$= 1 \, .$$

That is what we wished to prove. □

Theorem 3.34 (Minkowski's Inequality)   *Let f, g be functions in $L^p(\mathcal{R})$. Let $1 \leq p < \infty$. Then*

$$\int_{\mathcal{R}} |f(x) + g(x)|^p \, dx^{1/p} \leq \int_{\mathcal{R}} |f(x)|^p \, dx^{1/p} + \int_{\mathcal{R}} |g(x)|^p \, dx^{1/p} \, .$$

In other words,

$$\|f + g\|_{L^p} \leq \|f\|_{L^p} + \|g\|_{L^p} \, .$$

Proof. We shall use Hölder's inequality to good effect here.

Since $h(x) = x^p$ is a convex function, we know that

$$\left(\frac{1}{2} a + \frac{1}{2} b \right)^p \leq \frac{1}{2} a^p + \frac{1}{2} b^p \, .$$

Therefore,

$$(a + b)^p \leq 2^{p-1}a^p + 2^{p-1}b^p \,.$$

If $\|f + g\|_{L^p} = 0$, then Minkowski's inequality is trivially satisfied. So we may assume that $\|f + g\|_{L^p} \neq 0$. Thus,

$$
\begin{aligned}
\int_{\mathcal{R}} |f(x) + g(x)|^p \, dx \quad &\leq \quad \int_{\mathcal{R}} (|f(x)| + |g(x)|) \cdot |f(x) + g(x)|^{p-1} \, dx \\
&= \quad \int_{\mathcal{R}} |f(x)| \cdot |f(x) + g(x)|^{p-1} \, dx \\
&\quad + \int_{\mathcal{R}} |g(x)| \cdot |f(x) + g(x)|^{p-1} \, dx \\
&\overset{\text{(Hölder)}}{\leq} \quad \int_{\mathcal{R}} |f(x)|^p \, dx^{1/p} \cdot \int_{\mathcal{R}} |f(x) + g(x)|^p \, dx^{(p-1)/p} \\
&\quad + \int_{\mathcal{R}} |g(x)|^p \, dx^{1/p} \cdot \int_{\mathcal{R}} |f(x) + g(x)|^p \, dx^{(p-1)/p} \,.
\end{aligned}
$$

Now we may divide both sides of this inequality by $\int_{\mathcal{R}} |f(x) + g(x)|^p \, dx^{(p-1)/p}$ to obtain

$$\int_{\mathcal{R}} |f(x) + g(x)|^p \, dx^{1/p} \leq \int_{\mathcal{R}} |f(x)|^p \, dx^{1/p} + \int_{\mathcal{R}} |g(x)|^p \, dx^{1/p} \,.$$

In other words,

$$\|f + g\|_{L^p} \leq \|f\|_{L^p} + \|g\|_{L^p} \,,$$

as was to be proved. $\square$

The first of these two theorems is a useful device for estimate the inner product of two functions. The second of these results gives a triangle inequality on the Banach space L^p.

3.6   THE RIESZ REPRESENTATION THEOREM

A natural Banach space to study is the collection of continuous functions on a bounded interval $[a, b]$. Under the supremum norm, this is certainly a Banach space. What is its dual? The Riesz representation theorem gives an elegant answer to this question.

We define a *regular Borel measure* to be a real-valued function μ on the σ-algebra of Borel sets, which is countably subadditive and which satisfies these conditions:

(i) If E is a Borel set then

$$\mu(E) = \inf_{U \supseteq E} \mu(U) \,,$$

where the infimum here is taken over all open sets U that contain E.

(ii) The measure of any compact set is finite.

We also refer to such a measure as a *Radon measure*.

Now we have:

Theorem 3.35   *Let φ be a continuous linear functional on $C[a, b]$. Then there is a regular Borel measure μ on $[a, b]$ such that*

$$\varphi(f) = \int_{\mathcal{R}} f(x)\, d\mu(x)$$

for all $f \in C([a, b])$.

We refer to [RUD] for a particularly elegant proof due to John von Neumann (1903–1957).

3.7   PRODUCT INTEGRATION: FUBINI'S THEOREM

The theory of product integrals is rather technical, and we shall content ourselves here with the statements of some results and some illustrative discussion. No proofs will be provided.

We are interested in when a function on $\mathcal{R} \times \mathcal{R}$, or a suitable subset of $\mathcal{R} \times \mathcal{R}$, can be integrated. The first task is to discuss product measurable sets and product measure. Along the way, we shall talk about product Borel sets. In the end, we treat product integrals.

With $\mathcal{M}$ the σ-algebra of measurable sets on $\mathcal{R}$, we consider the σ-algebra on $\mathcal{R} \times \mathcal{R}$ generated by products $M \times N$ of elements of $\mathcal{M}$. Following Folland [FOL], we shall denote this σ-algebra by $\mathcal{M} \otimes \mathcal{M}$.

If $M, N \in \mathcal{M}$, then the set $M \times N$ is called a *rectangle*. If X is a finite, pairwise disjoint union of rectangles $M_j \times N_j$, then we may set

$$\widetilde{m}(X) = \sum m(M_j) \cdot m(N_j)\,.$$

For us, a simple function s shall be a finite linear combination of characteristic functions of pairwise disjoint rectangles R_j:

$$s(x) = \sum_{j=1}^{k} c_j \cdot \chi_{R_j}\,.$$

Of course we define

$$\iint_{\mathcal{R}^2} s(x) \equiv \sum_j c_j \cdot \widetilde{m}(R_j)\,.$$

Now, if f is a nonnegative, measurable function on $\mathcal{R}^2$, we define

$$\iint_{\mathcal{R}^2} f\, d\widetilde{m} = \sup_{s \leq f} \iint_{\mathcal{R}^2} s\, d\widetilde{m}\,.$$

If f is any measurable function on $\mathcal{R}^2$, we define

$$f^+(x) = f(x) \cdot \chi_{\{x:f(x)>0\}}\,,$$

$$f^-(x) = -f(x) \cdot \chi_{\{x:f(x)<0\}}\,.$$

Then we have:

Definition 3.36   If f is a measurable function, then we set

$$I^+ = \iint_{\mathcal{R}^2} f^+(x)\,d\widetilde{m}$$

and

$$I^- = \iint_{\mathcal{R}^2} f^-(x)\,d\widetilde{m}\,.$$

Finally, we define

$$\iint_{\mathcal{R}^2} f(x)\,d\widetilde{m} = I^+ - I^-$$

provided that both I^+ and I^- are finite.

In case $I^+ = \infty$ and I^- is finite, then we say that

$$\iint_{\mathcal{R}^2} f(x)\,d\widetilde{m} = +\infty\,.$$

In case $I^- = \infty$ and I_+ is finite, then we say that

$$\iint_{\mathcal{R}^2} f(x)\,d\widetilde{m} = -\infty\,.$$

In case *both* I^+ and I^- are infinite, then the integral does not exist.

We refer the reader to [FOL] for the full chapter and verse on product measures. We have attempted here only to set the stage for the Fubini-Tonelli theorem.

Theorem 3.37   (Fubini-Tonelli)

(a) *If f is nonnegative and measurable on $\mathcal{R} \times \mathcal{R}$, then the functions $f^1(x) = \int_{\mathcal{R}} f(x,t)dm(t)$ and $f^2(y) = \int_{\mathcal{R}} f(s,y)dm(s)$ have integrals that make sense on $\mathcal{R}$ (they could take the value ∞) and*

$$\iint_{\mathcal{R}^2} f\,d\widetilde{m} = \int_{\mathcal{R}} f^1(x)\,dm(x) = \int_{\mathcal{R}} f^2(y)\,dm(y)\,. \qquad (\star)$$

(b) *If f is integrable on $\mathcal{R} \times \mathcal{R}$, then f^1 is integrable as a function of one variables and f^2 is integrable as a function of one variable. Moreover, the equation $(\star)$ holds.*

3.8   THREE PRINCIPLES OF LITTLEWOOD

John Edensor Littlewood (1885–1941) was one of the seminal contributors to analysis in the twentieth century. His collaboration with Godfrey Harold Hardy (1877–1947) is legendary. He formulated three guiding principles in analysis, which we now enunciate:

1. Every measurable set is nearly a finite union of intervals.

2. Every measurable function is nearly continuous.

3. Every pointwise convergent sequence is nearly uniformly convergent.

Of course the only way to make any sense of these three precepts is to make more precise what we mean by "nearly." We shall formulate all of these ideas precisely. We shall also prove some of them, and discuss the others in detail.

Principle 1: By the way that we define the outer measure m, every measurable set $S \subseteq \mathcal{R}$ is contained in a union of open intervals the sum of whose lengths is nearly $m(S)$. That is the substance of the first principle.

Principle 2: This edict is contained in the following result:

Theorem 3.38   *Let f be a measurable function on an interval $[a, b]$, and assume that the set on which f takes the values $\pm\infty$ has measure 0. Let $\epsilon > 0$. Then there is a step function σ and a continuous function φ such that*

$$|f - \sigma| < \epsilon \qquad and \qquad |f - \varphi| < \epsilon$$

except on a set of measure less than ϵ.

Proof. We sketch the proof.

Step 1: Setting

$$S_j = \{x \in [a, b] : |f(x)| > 2^{-j}\} \qquad \text{for } j \in \mathcal{Z},$$

we see that $[a, b] \subseteq \cup_j S_j$. Hence, given $\epsilon > 0$, we see that there is an $M > 0$ such that $|f| \leq M$ except on a set of measure $\epsilon/3$.

Step 2: Following the proof of Proposition 3.19, we see that there is a simple function s so that $|s(x) - f(x)| < \epsilon$ except where $|f| > M$.

Step 3: With the simple function s as in Step 2, we may find a step function σ so that $s = \sigma$ except on a set of measure $\epsilon/3$.

Step 4: With the step function σ as in Step 3, we may find a continuous function φ so that $\varphi = \sigma$ except on a set of measure $\epsilon/3$.   □

Principle 3: The last precept is addressed by a notable theorem of Egoroff:

Theorem 3.39 *Let $E \subseteq \mathcal{R}$ be a measurable set of finite measure. Let f_j, f be measurable functions on E so that $f_j \to f$ almost everywhere. Then, for every $\epsilon > 0$, there is a measurable set $X \subseteq E$ such that $m(X) < \epsilon$ and $f_j \to f$ uniformly on $E \setminus X$.*

Proof. By discarding a set of measure 0, we may suppose that $f_j \to f$ at every point of E. For $j, k \in \mathcal{N}$, we let

$$E_j(k) = \bigcup_{m=j}^{\infty} \left\{ x \in E : |f_m(x) - f(x)| \geq \frac{1}{k} \right\}.$$

For a fixed k, the sets $E_j(k)$ decrease as j increases. Also, $\cap_{j=1}^{\infty} E_j(k) = \emptyset$; since $m(E) < \infty$, we may conclude that $m(E_j(k)) \to 0$ as $j \to \infty$. Given $\epsilon > 0$ and $k \in \mathcal{N}$, select a j_k so large that $m(E_{j_k}(k)) < \epsilon \cdot 2^{-k}$ and let $E = \cup_{k=1}^{\infty} E_{j_k}(k)$. Then $m(E) < \epsilon$, and we see that

$$|f_j(x) - f(x)| < \frac{1}{k} \qquad \text{for } j > j_k \text{ and } x \notin X.$$

We conclude that $f_j \to f$ on $E \setminus X$. □

3.9 DIFFERENTIATION OF INTEGRALS: COVERING LEMMAS AND THE LEBESGUE THEOREM

3.9.1 BASIC IDEAS

Our presentation here owes a debt to [KRP].

Let $S \subseteq \mathcal{R}$ be a set. A *covering* of S shall be a collection $\mathcal{U} = \{U_\alpha\}_{\alpha \in \mathcal{A}}$ of sets such that $\bigcup_{\alpha \in \mathcal{A}} U_\alpha \supseteq S$. If all the sets of $\mathcal{U}$ are open, then we call $\mathcal{U}$ an *open covering* of S. A *subcovering* of the covering $\mathcal{U}$ is a covering $\mathcal{V} = \{V_\beta\}_{\beta \in \mathcal{B}}$ such that each V_β is one of the U_α. A *refinement* of the covering $\mathcal{U}$ is a collection $\mathcal{W} = \{W_\gamma\}_{\gamma \in \mathcal{G}}$ of sets such that each W_γ is a subset of some U_α and $\{W_\gamma\}$ is still a covering of S. If $\mathcal{U}$ is a covering of a set S then the *valence* of $\mathcal{U}$ is the least positive integer M such that no point of S lies in more than M of the sets in $\mathcal{U}$.

It is easy to see that any open cover of a set $S \subseteq \mathcal{R}$ has a countable subcover (see [HUW]). We also know (Lebesgue) that any open covering of S has a refinement with valence at most 2 (see [HUW, Theorem V 1]).

Wiener's covering lemma concerns a covering of a set by a collection of intervals. Let us use the notation $I(p, r)$ to denote $\{x \in \mathcal{R} : |x - p| < r\}$— in other words, the interval with center p and "radius" r.

Lemma 3.40 (Wiener) *Let $K \subseteq \mathcal{R}$ be a compact set with a covering $\mathcal{U} = \{I_\alpha\}_{\alpha \in A}$, $I_\alpha = I(c_\alpha, r_\alpha)$, by open intervals. Then there is a subcollection $I_{\alpha_1}, I_{\alpha_2}, \ldots, I_{\alpha_m}$, consisting of pairwise pairwise disjoint* *intervals, such that*

$$\bigcup_{j=1}^{m} (c_{\alpha_j}, 3r_{\alpha_j}) \supseteq K.$$

Proof. Since K is compact, we may assume that there are only finitely many I_α. Let I_{α_1} be the interval in this collection that has the greatest radius (this interval may not be unique). Let I_{α_2} be the interval that is pairwise disjoint from I_{α_1} and has greatest radius among those intervals that are pairwise disjoint from I_{α_1} (again, this interval may not be unique). At the jth step, choose the (not necessarily unique) interval pairwise disjoint from $I_{\alpha_1}, \ldots, I_{\alpha_{j-1}}$ that has greatest radius among those intervals that are pairwise disjoint from $I_{\alpha_1}, \ldots, I_{\alpha_{j-1}}$. Continue. The process ends in finitely many steps. We claim that the I_{α_j} chosen in this fashion do the job.

For each j, we shall write $I_{\alpha_j} = I(c_{\alpha_j}, r_{\alpha_j})$. It is enough to show that $I_\alpha \subseteq \bigcup_j I(c_{\alpha_j}, 3r_{\alpha_j})$ for every α. Fix an α. If $\alpha = \alpha_j$ for some j, then we are done. If $\alpha \notin \{\alpha_j\}$, let j_0 be the first index j with $I_{\alpha_j} \cap I_\alpha \neq \emptyset$ (there must be one; otherwise the process would not have stopped). Then $r_{\alpha_{j_0}} \geq r_\alpha$; otherwise we selected $I_{\alpha_{j_0}}$ incorrectly. But then (by the triangle inequality) $(c_{\alpha_{j_0}}, 3r_{\alpha_{j_0}}) \supseteq (c_\alpha, r_\alpha)$ as desired. $\square$

3.9.2   THE MAXIMAL FUNCTION

A classical idea, due to Hardy and Littlewood, is the so-called maximal function. It is used to control other operators, and also to study questions of differentiation of integrals. The covering lemma from the previous section is used to control these new maximal functions.

Definition 3.41   If f is a locally integrable function on $\mathcal{R}$, then we let $Mf(x)$ be given by

$$Mf(x) = \sup_{R > 0} \frac{1}{m[I(x, R)]} \int_{I(x,R)} |f(t)| \, dm(t).$$

The operator M is called the *Hardy–Littlewood maximal operator*. A few facts are immediately obvious about M:

1. M is not linear, but it is *sublinear* in the sense that

$$M[f + g](x) \leq Mf(x) + Mg(x).$$

2. Mf is always non-negative, and it *could* be identically equal to infinity.

3. Mf makes sense for any locally integrable f.

We shall, in fact, prove that Mf is finite m-almost everywhere, for any $f \in L^p$. In order to do so, it is convenient to formulate a weak notion of boundedness for operators. This idea shall be strictly milder than ordinary boundedness on L^p, but it interacts with that more classical notion in a productive fashion. The idea is due to M. Riesz.

To begin, we say that a measurable function f is *weak type* p, $1 \leq p < \infty$, if there exists a $C = C(f)$ with $0 < C < \infty$ such that, for any $\lambda > 0$,

$$m(\{x \in \mathcal{R} : |f(x)| > \lambda\}) \leq \frac{C}{\lambda^p}.$$

An operator T on L^p, taking values in the collection of measurable functions, is said to be of *weak type* (p, p) if there exists a constant $C = C(T)$ with $0 < C < \infty$ such that, for any $f \in L^p$ and for any $\lambda > 0$,

$$m(\{x \in \mathcal{R} : |Tf(x)| > \lambda\}) \leq C \cdot \left(\frac{\|f\|_{L^p}}{\lambda} \right)^p.$$

A function is defined to be *weak type* ∞ when it is L^∞. For $1 \leq p < \infty$, an L^p function is certainly weak type p, but the converse is not true. In fact, we note that the function $f(x) = |x|^{-1/p}$ on $\mathcal{R}^1$ is weak type p, but not in L^p, for $1 \leq p < \infty$. The Hilbert transform (see [KRA1]) is an important operator that is not bounded on L^1 but is in fact weak type $(1, 1)$.

Proposition 3.42   *The Hardy–Littlewood maximal operator M is weak type $(1, 1)$.*

Proof. Let $\lambda > 0$. Set $S_\lambda = \{x : |Mf(x)| > \lambda\}$. Because Mf is the supremum of a set of continuous functions, Mf is lower semicontinuous and, consequently, S_λ is open.

Since S_λ is open, we may let $K \subseteq S_\lambda$ be a compact subset with $2\,m(K) \geq m(S_\lambda)$. For each $x \in K$, there is a interval $I_x = I(x, r_x)$ with

$$\lambda < \frac{1}{m(I_x)} \int_{I_x} |f(t)|dm(t).$$

Then $\{I_x\}_{x \in K}$ is an open cover of K by intervals. By Lemma 3.40, there is a subcollection $\{I_{x_j}\}_{j=1}^M$ which is pairwise pairwise disjoint, but so that the threefold dilates of these selected intervals still covers K. Then

$$m(S_\lambda) \leq 2\,m(K) \quad \leq \quad 2\,m\left(\bigcup_{j=1}^M I(x_j, 3r_j) \right)$$

$$\leq 2 \sum_{j=1}^M m[I(x_j, 3r_j)]$$

$$\leq \sum_{j=1}^{M} 2 \cdot 3 \, m(I_{x_j})$$

$$\leq \sum_{j=1}^{M} \frac{2 \cdot 3}{\lambda} \int_{I_{x_j}} |f(t)| dm(t)$$

$$\leq \frac{6}{\lambda} \|f\|_{L^1} . \qquad \square$$

It is completely obvious that the Hardy-Littlewood maximal operator M is bounded (with norm 1) on the space L^∞. It follows then from the Marcinkiewicz interpolation theorem [KRA2, STE]) that M is bounded on L^p, $1 < p < \infty$. We shall not need this last result right away, but it is important for our understanding of these operators. $\qquad \square$

One of the venerable applications of the Hardy–Littlewood operator is the Lebesgue differentiation theorem:

Theorem 3.43 *Let f be a locally Lebesgue integrable function on $\mathcal{R}$. Then, for m–almost every $x \in \mathcal{R}$, it holds that*

$$\lim_{R \to 0^+} \frac{1}{m[I(x, R)]} \int_{I(x,R)} f(t) \, dm(t) = f(x) .$$

Proof. Multiplying f by a compactly supported C^∞ function that is identically 1 on an interval, we may as well suppose that $f \in L^1$. We begin by proving that

$$\lim_{R \to 0^+} \frac{1}{m[(x, R)]} \int_{(x,R)} f(t) \, dm(t)$$

exists.

Let $\epsilon > 0$. Select a function φ, continuous with compact support, and real-valued, so that $\|f - \varphi\|_{L^1} < \epsilon^2$. Then

$$m\left\{ x \in \mathcal{R} : \left| \limsup_{R \to 0^+} \frac{1}{m[I(x, R)]} \int_{I(x,R)} f(t) \, dm(t) \right. \right.$$
$$\left. \left. - \liminf_{R \to 0^+} \frac{1}{m[I(x, R)]} \int_{I(x,R)} f(t) \, dm(t) \right| > \epsilon \right\}$$

$$\leq \ m\left\{x \in \mathcal{R} : \limsup_{R \to 0^+} \frac{1}{m[I(x, R)]} \int_{I(x,R)} |f(t) - \varphi(t)| \, dm(t) \right.$$

$$+ \left| \limsup_{R \to 0^+} \frac{1}{m[I(x, R)]} \int_{I(x,R)} \varphi(t) \, dm(t) \right.$$

$$- \liminf_{R \to 0^+} \frac{1}{m[I(x, R)]} \int_{I(x,R)} \varphi(t) \, dm(t) \Big|$$

$$\left. + \limsup_{R \to 0^+} \frac{1}{m[I(x, R)]} \int_{I(x,R)} |\varphi(t) - f(t)| \, dm(t) > \epsilon \right\}$$

$$\leq \ m\left\{x \in \mathcal{R} : \limsup_{R \to 0^+} \frac{1}{m[I(x, R)]} \int_{I(x,R)} |f(t) - \varphi(t)| \, dm(t) > \frac{\epsilon}{3} \right\}$$

$$+ m\left\{x \in \mathcal{R} : \left| \limsup_{R \to 0^+} \frac{1}{m[I(x, R)]} \int_{I(x,R)} \varphi(t) \, dm(t) \right.\right.$$

$$\left.\left. - \liminf_{R \to 0^+} \frac{1}{m[I(x, R)]} \int_{I(x,R)} \varphi(t) \, dm(t) \right| > \frac{\epsilon}{3} \right\}$$

$$+ m\left\{x \in \mathcal{R} : \limsup_{R \to 0^+} \frac{1}{m[I(x, R)]} \int_{I(x,R)} |\varphi(t) - f(t)| \, dm(t) > \frac{\epsilon}{3} \right\}$$

$$\equiv \ I + II + III \,.$$

Now $II = 0$ because the set being measured is empty (since φ is continuous). Each of I and III may be estimated by

$$m\left\{x \in \mathcal{R} : M(f - \varphi)(x) > \epsilon/3 \right\}$$

and this, by Proposition 3.42, is majorized by

$$C \cdot \frac{\epsilon^2}{\epsilon/3} = c \cdot \epsilon \,.$$

In sum, we have proved the estimate

$$m\left\{x \in \mathcal{R} : \left| \limsup_{R \to 0^+} \frac{1}{m[I(x, R)]} \int_{I(x,R)} f(t) \, dm(t) \right.\right.$$

$$\left.\left. - \liminf_{R \to 0^+} \frac{1}{m[I(x, R)]} \int_{I(x,R)} f(t) \, dm(t) \right| > \epsilon \right\} \leq c \cdot \epsilon \,.$$

It follows immediately that

$$\lim_{R \to 0^+} \frac{1}{m[I(x, R)]} \int_{I(x,R)} f(t) \, dm(t)$$

exists for m-almost every $x \in \mathcal{R}$.

The proof that the limit actually equals $f(x)$ at m-almost every point follows exactly the same lines. We shall omit the details.   □

Corollary 3.44   *If $A \subseteq \mathcal{R}$ is Lebesgue measurable, then, for almost every $x \in \mathcal{R}$, it holds that*

$$\lim_{r \to 0^+} \frac{m(A \cap I(x, r))}{m(I(x, r))} = 1 \,.$$

Proof. Set $f = \chi_A$. Then

$$\int_{I(x,R)} f(t) \, dm(t) = m(A \cap I(x, r))$$

and the corollary follows from Theorem 3.43.   □

A point for which this last corollary is true is called a *point of density* or a *Lebesgue point* of the set A.

It is possible to construe the Hardy–Littlewood maximal function in the more general context of measures.

Definition 3.45   Let μ be a Radon measure on $\mathcal{R}$. If f is a μ-measurable function and $x \in \mathcal{R}$, then we define

$$M_\mu f(x) = \sup_{r > 0} \frac{1}{\mu[I(x, r)]} \int_{I(x,r)} |f(t)| \, d\mu(t) \,.$$

Further, and more generally, if ν is a Radon measure on $\mathcal{R}$, then we define

$$M_\mu \nu(x) = \sup_{r > 0} \frac{\nu[(x, r)]}{\mu[(x, r)]} \,.$$

Finally, it is sometimes useful to have the non-centered maximal operator $\widetilde{M}_\mu$ defined by

$$\widetilde{M}_\mu f(x) = \sup_{I(z,r) \ni x} \frac{1}{\mu[I(z, r)]} \int_{I(z,r)} |f(t)| \, d\mu(t) \,.$$

A similar definition may be given for the maximal function of a Radon measure.

It is not difficult to see that the non-centered maximal operator $\widetilde{M}$ is comparable to the centered one M. But it is useful, for the sake of flexibility, to have both operators at our disposal.

The principal result about these maximal functions is the following:

Theorem 3.46   *The operator M_μ is weak type $(1, 1)$ in the sense that*

$$\mu \left\{ x \in \mathcal{R} : M_\mu \nu(x) > s \right\} \le C \cdot \frac{\nu(\mathcal{R})}{s} \,.$$

In particular, if $f \in L^1(\mu)$, then

$$\mu\left\{x \in \mathcal{R} : M_\mu f(x) > s\right\} \leq C \cdot \frac{\|f\|_{L^1}}{s} .$$

In case the measure μ satisfies the dilation condition $\mu[I(x, 3r)] \leq c \cdot \mu[I(x, r)]$, then we have

$$\mu\left\{x \in \mathcal{R} : \widetilde{M}_\mu v(x) > s\right\} \leq c \cdot s^{-1} \cdot v\left\{x \in \mathcal{R} : \widetilde{M}_\mu v(x) > s\right\} .$$

The proof of this result follows the same lines as the development of Proposition 3.42 and we omit the details. A full account may be found in [MAT]. It also follows, as before, that M_μ and $\widetilde{M}_\mu$ are bounded on L^∞. Hence, by interpolation, they are bounded on L^p for $1 < p < \infty$.

3.10   THE CONCEPT OF CONVERGENCE IN MEASURE

Whereas, on a finite-dimensional space, all natural topologies tend to be equivalent, on an infinite-dimensional space there are many different topologies. The space of measurable functions is such a space. We have already seen the topology of pointwise convergence. We have also seen the L^p-space topology. Now we shall take a look at the very useful topology given by the concept of convergence in measure.

Let f_j, f be measurable functions on a measurable set $E \subseteq \mathcal{R}$. We say that the f_j *converge in measure* to f if, for each $\epsilon > 0$,

$$m\left(\{x \in \mathcal{R} : |f_j(x) - f(x)| > \epsilon\}\right) \to 0 \quad \text{as } j \to \infty .$$

In a similar vein, a sequence $\{f_j\}$ is *Cauchy in measure* if, for each $\epsilon > 0$,

$$m\left(\{x : |f_j(x) - f_k(x)| > \epsilon\}\right) \to 0 \quad \text{as } j, k \to \infty .$$

Example 3.47   Let $f_j(x) = 2^{-j} \cdot \chi_{(0,j)}(x)$. Then it is plain that $f_j \to 0$ in measure.

Example 3.48   Let $f_j(x) = \chi_{(j,j+1)}(x)$. This sequence is *not* Cauchy in measure.

Example 3.49   Let $g_j(x) = 2^j \cdot \chi_{[0,1/j]}$. Then $g_j \to 0$ in measure.

Example 3.50   Let $h_p(x) = \chi_{[j \cdot 2^{-k}, (j+1) \cdot 2^{-k}]}$, where $0 \leq j < 2^k$ and $p = j + 2^k$. Thus, h_p is the characteristic function of a subinterval of the unit interval having length 2^{-k}. Then $h_p \to 0$ in measure.

Proposition 3.51   *Let $\{f_j\}$ be a sequence of measurable functions that is Cauchy in measure. Then there is a measurable function f such that $f_j \to f$ in measure. Also there is a subsequence f_{j_k} which converges to f almost everywhere.*

Proof. Select a subsequence $\{g_k\} \equiv \{f_{j_k}\}$ of $\{f_j\}$ such that, if $F_k = \{x \in \mathcal{R} : |g_k(x) - g_{k+1}(x)| > 2^{-k}\}$, then $m(F_k) \leq 2^{-k}$. If $G_m = \cup_{k=m}^{\infty} F_k$, then $m(G_m) \leq \sum_{k=m}^{\infty} 2^{-k} = 2^{1-m}$. If $x \notin G_m$, then, for $\ell \geq k \geq m$, we have

$$|g_k(x) - g_\ell(x)| \leq \sum_{p=k}^{\ell-1} |g_{p+1}(x) - g_p(x)| \leq \sum_{p=k}^{\ell-1} 2^{-p} \leq 2^{1-k}. \tag{$\dagger$}$$

Hence, $\{g_k\}$ is pointwise Cauchy on ${}^c G_m$.

Let $G = \cap_{m=1}^{\infty} G_m = \limsup_{k \to \infty} F_k$. Then $m(G) = 0$. Also, if we set $g(x) = \lim_{k \to \infty} g_k(x)$ for $x \notin G$ and $g(x) = 0$ for $x \in G$, then g is measurable and $g_k \to g$ almost everywhere. Also line ($\dagger$) shows that $|g_k(x) - g(x)| \leq 2^{2-k}$ for $k \geq m$ and $x \notin G_m$. Since $m(G_m) \to 0$ as $m \to \infty$, we see that $g_k \to g$ in measure.

But then $f_n \to g$ in measure because

$$\{x \in \mathcal{R} : |f_n(x) - g(x)| > \epsilon\}$$
$$\subseteq \{x \in \mathcal{R} : |f_n(x) - g_k(x)| > \epsilon/2\} \cup \{x \in \mathcal{R} : |g_k(x) - g(x)| > \epsilon/2\}.$$

The sets on the right both have small measure when n and k are large. $\square$

3.11 FUNCTIONS OF BOUNDED VARIATION AND ABSOLUTE CONTINUITY

If f is an integrable function (i.e., an L^1 function) on $\mathcal{R}$, then it is natural to form the new function

$$F(x) = \int_a^x f(t)\, dt$$

and to ask under what circumstances we may say that the derivative $F'(x)$ exists and equals $f(x)$. In the context of the Riemann integral, we have a positive answer when f is continuous. For the Lebesgue integral, the answer is different and a bit more sophisticated.

In this section, we sketch the answer to this question. We shall not give full proofs of all statements, but we shall paint a rather complete picture.

We begin with some definitions:

Definition 3.52   Let $\mathcal{M}$ be the usual collection of measurable sets in $\mathcal{R}$. A *signed measure* is a function $\nu : \mathcal{M} \to \overline{\mathcal{R}}$ such that

(i) $v(\emptyset) = 0$.

(ii) v does not assume both values $+\infty$ and $-\infty$.

(iii) If $\{E_j\}$ is a sequence of pairwise pairwise disjoint sets in $\mathcal{R}$ then $m(\cup_j E_j) = \sum_j m(E_j)$.

A signed measure taking values in $[0, \infty]$ is called a *positive measure*.

Example 3.53   Let f be an integrable function on $\mathcal{R}$. Define

$$v(E) = \int_E f(x) \, dm(x) \,.$$

Then v is a signed measure.

Example 3.54   Lebesgue measure is a positive measure.

Definition 3.55   Let E be a measurable set. Let μ and v be signed measures. We say that μ and v are *mutually singular*, and we write $\mu \perp v$, if we can write $E = X \cup Y$, $X \cap Y = \emptyset$, $\mu(X) = 0$, and $v(Y) = 0$.

It is clear that μ and v are mutually singular if they "live on pairwise disjoint sets."

Definition 3.56   Suppose that v is a signed measure and μ is a positive measure. We say that v is *absolutely continuous* with respect to μ, and we write $v << \mu$ if, for each $\epsilon > 0$, there is a $\delta > 0$ such that, when $\mu(E) < \delta$, then $v(E) < \epsilon$.

Clearly, if $v << \mu$ and $\mu(E) = 0$, then $v(E) = 0$.

Definition 3.57   Let us call a measure μ by the name σ-*finite* if we can write $\mathcal{R} = \cup_j E_j$ and the measure of each E_j is finite.

Theorem 3.58   (Lebesgue-Radon-Nikodym)   *If v is a signed measure and μ is a σ-finite measure, then there exists a unique measure λ and an $f \in L^1(\mathcal{R})$ so that $\lambda \perp \mu$ and $dv = d\lambda + f d\mu$.*

This is a fundamental result of measure theory. But its proof is complicated and technical, and we shall omit it.

Definition 3.59   A Borel measure μ is called *regular* if

(a) The μ-measure of any compact set is finite.

(b) The measure of any set is the infimum of measures of open sets that contain it.

In what follows, $I(x, r)$ shall denote a Euclidean interval with center x and radius r.

Theorem 3.60   *Let v be a regular, signed, Borel measure. Let $dv = d\lambda + f\,dm$ be its Lebesgue-Radon-Nokodym decomposition. In particular, $\lambda \perp m$ and $f \in L^1$. Then, for m almost every $x \in \mathcal{R}$,*

$$\lim_{r \to 0} \frac{v(I(x, r))}{m(I(x, r))} = f(x).$$

Proof. Certainly the measure $f\,dm$ is regular so λ is regular as well. In view of the Lebesgue differentiation theorem (Theorem 3.43 of Section 3.9), it suffices for us to show that, if λ is regular and $\lambda \perp m$, then

$$\frac{\lambda(I(x, r))}{m(I(x, r))} \to 0$$

for m almost every x. We may assume that λ is positive.

Let A be a Borel set such that $\lambda(A) = m({}^c A) = 0$. Set

$$F_k = \left\{ x \in A : \limsup_{r \to 0} \frac{\lambda(I(x, r))}{m(I(x, r))} > \frac{1}{k} \right\}.$$

We show that $m(F_k) = 0$ for all k. That establishes our result.

The argument that we now present is familiar from our study of the Hardy-Littlewood maximal function. Let $\epsilon > 0$. Since λ is regular, there is an open set $U_\epsilon \supseteq A$ such that $\lambda(U_\epsilon) < \epsilon$. Each $x \in F_k$ is the center of an interval $I(x, r_x) \equiv I_x$ such that $I_x \subseteq U_\epsilon$ and $\lambda(I_x) > (1/k) \cdot m(I_x)$. Now the covering Lemma 3.40 tells us that if $V_\epsilon = \cup_{x \in F_k} I_x$ and $c < m(V_\epsilon)$ then there exist $x_1, \ldots x_k$ such that $I_{x_1}, I_{x_2}, \ldots, I_{x_k}$ are pairwise pairwise disjoint and

$$c < 3 \sum_{j=1}^{k} m(I_{x_j}) \leq 3k \sum_{j=1}^{k} \lambda(I_{x_j}) \leq 3k\lambda(V_\epsilon) \leq 3k\lambda(U_\epsilon) \leq 3k\epsilon.$$

We see that $M(V_\epsilon) \leq 3k\epsilon$. Since $F_k \subseteq V_\epsilon$ and $\epsilon > 0$ was arbitrary, we conclude that $m(F_k) = 0$. $\square$

Definition 3.61   Let $f : \mathcal{R} \to \mathcal{R}$. Set

$$F_f(x) = \sup \left\{ \sum_{j=1}^{k} |f(x_j) - f(x_{j-1})| : -\infty < x_0 < x_1 < \cdots x_k = x \right\}.$$

We call T_f the *total variation function* of f. In case $T_f(\infty) = \lim_{x \to \infty} T_f(x)$ is finite, then we say that the function f is of *bounded variation*, and we write $f \in BV$. If $f(-\infty) = 0$ and f is right continuous, we say that f is of *normalized bounded variation*, and we write $f \in NBV$.

Theorem 3.62   *If μ is a signed Borel measure on $\mathcal{R}$ and $f(x) \equiv \mu((-\infty, x])$, then $f \in NBV$. Conversely, if $f \in NBV$, then there is a unique signed measure μ_f such that $f(x) = \mu_f((-\infty, x])$.*

Proof. We may write $\mu = \mu^+ - \mu^-$, where μ^+ and μ^- are nonnegative measures (this is called the *Jordan decomposition* of μ—see [FOL]). Set

$$f^+(x) = \mu^+((-\infty, x]) \qquad \text{and} \qquad f^-(x) = \mu^-((-\infty, x]).$$

We see that f^+, f^- are increasing and right continuous, and that $f^+(-\infty) = 0$, $f^-(-\infty) = 0$. It follows that $f = f^+ - f^-$ is in NBV.

Conversely, any $f \in NBV$ can be written in the indicated form (simply define $\mu((-\infty, x])$ to be $f(x)$) and μ may be decomposed as the difference of two positive measures. $\square$

Proposition 3.63   *If $f \in NBV$ then $f' \in L^1$. Furthermore, $\mu_f \perp m$ if and only if $f' = 0$ almost everywhere. Finally, $\mu_f << m$ if and only if $f(x) = \int_{-\infty}^x f'(t)\,dt$.*

Proof. Merely note that

$$f'(x) = \lim_{r \to 0} \frac{\mu_f(E_r)}{m(E_r)},$$

where E_r is the half-interval $(x, x+r]$. Now apply Theorem 3.60 above. $\square$

Definition 3.64   A function $f : \mathcal{R} \to \mathcal{R}$ is said to be *absolutely continuous* if, whenever $\epsilon > 0$, then there is a $\delta > 0$ so that if $(a_1, b_1), (a_2, b_2), \ldots (a_k, b_k)$ are pairwise disjoint intervals in $\mathcal{R}$ with $\sum_j (b_j - a_j) < \delta$, then

$$\sum_j |f(b_j) - f(a_j)| < \epsilon.$$

It is easy to see that the condition for absolute continuity is stronger than that for bounded variation.

Proposition 3.65   *Let $f \in NBV$. Then f is absolutely continuous if and only if $\mu_f << m$.*

Proof. If $\mu_f << m$, then the absolute continuity of f follows immediately from the definitions.

For the converse, suppose that E is a Borel set with $m(E) = 0$. If ϵ and δ are as in the definition of absolute continuity, then we can find open sets $U_1, U_2, \ldots$ such that $U_j \supseteq U_{j+1}$ for each j and each U_j contains E and so that $m(U_1) < \delta$ and $\mu_f(U_j) \to \mu_f(E)$. Each U_j is a pairwise disjoint union of open intervals (a_ℓ^j, b_ℓ^j) and

$$\sum_{\ell=1}^{N} |\mu_f((a_\ell^j, b_\ell^j))| \leq \sum_{\ell=1}^{N} |f(b_\ell^j) - f(a_\ell^j)| < \epsilon$$

for all N. Letting $N \to \infty$, we see that $|\mu_f(U_j)| < \epsilon$, hence $|\mu_f(E)| \leq \epsilon$. Since ϵ was chosen arbitrarily, we conclude that $\mu_f(E) = 0$. Therefore, $\mu_f << m$. $\qquad\square$

Corollary 3.66   *If $f \in L^1$, then the function*

$$F(x) = \int_{-\infty}^{x} f(t)\, dt$$

is in NBV and is absolutely continuous. Furthermore, $F' = f$ almost everywhere. Conversely, if $F \in NBV$ is absolutely continuous, then $F' \in L^1$ and

$$F(x) = \int_{-\infty}^{x} F'(t)\, dt .$$

Proof. This is immediate from the preceding two results. $\qquad\square$

Corollary 3.67   *If f is absolutely continuous on $[a, b]$, then f is of bounded variation on $[a, b]$.*

Proof. This is almost immediate from the definitions. We leave the details as an exercise. $\qquad\square$

And now we come to the main result of this body of material:

Theorem 3.68   (The Fundamental Theorem of Calculus)   *If $-\infty < a < b < \infty$ and $f : [a, b] \to \mathcal{R}$, then the following are equivalent:*

(a) *f is absolutely continuous on $[a, b]$.*

(b) *$f(x) - f(a) = \int_a^x \varphi(t)\, dt$ for some $\varphi \in L^1([a, b])$.*

(c) *f is differentiable almost everywhere on $[a, b[$, $f' \in L^1([a, b])$, and $f(x) - f(a) = \int_a^x f'(t)\, dt$.*

Proof. In proving that **(a)** implies **(c)**, we may suppose that $f(a) = 0$. Set $f(x) = 0$ for $x < a$ and $f(x) = f(b)$ for $x > b$. Then $f \in NBV$ by the preceding result. So **(c)** follows from Corollary 3.66.

That **(c)** implies **(b)** is obvious.

Finally, we may see that **(b)** implies **(a)** by setting $f(t) = 0$ for $t \notin [a, b]$ and applying the corollary above.   □

EXERCISES

1. Prove that, if f is a continuously differentiable function on the interval $[a, b]$, then

$$Vf = \int_a^b |f'(x)| \, dx .$$

[*Hint:* You will prove two inequalities. For one, use the Fundamental Theorem. For the other, use the Mean Value Theorem.]

2. Let $f(x) = \sin x$ on the interval $[0, 2\pi]$. Calculate $Vf(x)$ for any $x \in [0, 2\pi]$.

3. Give an example of a continuous function on the interval $[0, 1]$ that is not of bounded variation.

4. Define $f_j(x) = \chi_{[j, j+1]}$. Discuss what the Lebesgue monotone convergence theorem, the Lebesgue dominated convergence theorem, and Fatou's lemma have to say about this sequence of functions. [This sequence is affectionately known as the "gliding hump."]

5. Define $g_j(x) = j \cdot \chi_{[1-1/j, 1]}$. Discuss what the Lebesgue monotone convergence theorem, the Lebesgue dominated convergence theorem, and Fatou's lemma have to say about this sequence of functions.

6. Define $h_j(x) = \chi_{[1-1/j, 1]}$. Discuss what the Lebesgue monotone convergence theorem, the Lebesgue dominated convergence theorem, and Fatou's lemma have to say about this sequence of functions.

7. Let f be an integrable function on the real line. Prove that, for any $\lambda > 0$, we have the inequality

$$m(\{x \in \mathcal{R} : |f(x)| > \lambda\}) \leq \frac{\|f\|_{L^1}}{\lambda} .$$

8. Refer to Exercise 7. Let f be a pth-power integrable function on the real line. So $|f|^p$ is integrable. Prove that, for any $\lambda > 0$, we have the inequality

$$m(\{x \in \mathcal{R} : |f(x)| > \lambda\}) \leq \left(\frac{\|f\|_{L^p}}{\lambda}\right)^p .$$

9. Let f_j, f be measurable functions on $\mathcal{R}$. We say that $f_j \to f$ in the L^p topology if

$$\lim_{j \to \infty} \int_{\mathcal{R}} |f_j(x) - f(x)|^p \, dx^{1/p} = 0.$$

Prove that, if $f_j \to f$ in the L^p topology, then

$$\|f_j\|_{L^p} \to \|f\|_{L^p}$$

as $j \to \infty$.

10. Refer to Exercise 9. Let f_j, f be bounded, measurable functions. Assume that $f_j \to f$ uniformly on compact sets. For which p is it true that f_j converges to f in the L^p topology?

11. Let f be an integrable function on $\mathcal{R}$. Show that, if $\epsilon > 0$, then there is a $\delta > 0$ so that if $m(E) < \delta$, then

$$\int_E |f(x)| \, dx < \epsilon.$$

12. Prove the Riemann-Lebesgue lemma: If f is integrable on $\mathcal{R}$, then

$$\lim_{j \to \infty} \int_{\mathcal{R}} f(x) \sin jx \, dx = 0.$$

[**Hint:** First prove the result for f continuously differentiable with compact support. Then approximate.]

13. Prove that if f is measurable and nonnegative and $\int_{\mathcal{R}} f(x) \, dx = 0$, then $f = 0$ almost everywhere.

14. Let f be an integrable function on $\mathcal{R}$. Prove that

$$\lim_{a \to 0} \int_{\mathcal{R}} |f(x + a) - f(x)| \, dx = 0.$$

[**Hint:** First prove the result when f is continuous with compact support. Then approximate.]

16. Show that we can have strict inequality in Fatou's lemma. [**Hint:** Think about the gliding hump, as in Exercise 4.]

17. Suppose that E is measurable and that E is a subset of the nonmeasurable set that we constructed at the beginning of the chapter. Show that $m(E) = 0$.

18. Let f be a nonnegative, measurable function on $\mathcal{R}$ that is integrable. Define

$$F(x) = \int_{-\infty}^{x} f(t) \, dt.$$

Prove that F is continuous at every point.

19. Show that there is no monotone convergence theorem for a decreasing sequence of functions.

21. Give an example of a function on $[0, \infty)$ that has a convergent improper Riemann integral but which is, in fact, *not* Lebesgue integrable.

22. Let g be an integrable function on $\mathcal{R}$ and suppose that $\{f_j\}$ is a sequence of measurable functions on $\mathcal{R}$ such that $|f_j| \le g$ for all j. Prove that

$$\int_{\mathcal{R}} \liminf_{j \to \infty} f_j(x)\, dx \le \liminf_{j \to \infty} \int_{\mathcal{R}} f_j(x)\, dx$$
$$\le \limsup_{j \to \infty} \int_{\mathcal{R}} f_j(x)\, dx$$
$$\le \int_{\mathcal{R}} \limsup_{j \to \infty} f_j(x)\, dx .$$

23. What are the weakest hypotheses on measurable f_j, f that imply that

$$\int_{\mathcal{R}} |f_j(x)|\, dx \to \int_{\mathcal{R}} |f(x)|\, dx ?$$

24. Let f be an integrable function on $\mathcal{R}$. Let $\epsilon > 0$. Show that there is a simple function s so that

$$\int_{\mathcal{R}} |f(x) - s(x)|\, dx < \epsilon .$$

25. Let f be an integrable function on $\mathcal{R}$. Let $\epsilon > 0$. Show that there is a continuous function φ so that

$$\int_{\mathcal{R}} |f(x) - \varphi(x)|\, dx < \epsilon .$$

26. Let f be a continuously differentiable function on $[0, 1] \times [0, 1]$. Show that

$$\frac{d}{dt} \int_0^1 f(s, t)\, ds = \int_0^1 \frac{\partial}{\partial t} f(x, t)\, ds .$$

CHAPTER 4

Comparison of the Riemann and Lebesgue Integrals

4.1 ANY RIEMANN INTEGRABLE FUNCTION IS LEBESGUE INTEGRABLE

Our theorem is this:

Theorem 4.1 *Let $[a, b] \subseteq \mathcal{R}$ be a bounded interval. If f is Riemann integrable on $[a, b]$, then f is Lebesgue integrable on $[a, b]$.*

Proof. Let $\epsilon > 0$. Let $\mathcal{P} = \{p_0, p_1, \ldots, p_k\}$ be a partition of $[a, b]$. In each interval $[p_{j-1}, p_j]$, select a point where $f(\xi_j)$ is within ϵ of the infimum of f on the interval. The corresponding Riemann sum

$$\sum_j f(\xi_j) \Delta_j$$

is, in fact, the Lebesgue integral of the simple function

$$\sum_j f(\xi_j) \chi_{[p_{j-1}, p_j]}. \tag{*}$$

These converge, as the partition becomes finer, by our hypothesis. But, by our observation regarding line ($*$), they converge to the Lebesgue integral of f as well. □

It is straightforward to check—merely by examining the proof just presented—that if f is Riemann integrable, then the value of its Riemann integral equals the value of its Lebesgue integral. Often we calculate the Lebesgue integral of a function by actually calculating its Riemann integral.

Example 4.2 Define

$$f(x) = \begin{cases} 1/q & \text{if} \quad x = p/q \text{ is a rational number in lowest terms} \\ 0 & \text{if} \quad x \text{ is irrational}. \end{cases}$$

Then it is straightforward to check that f is continuous at each irrational number and discontinuous at each rational number. We know that the set of rational numbers has measure 0. It follows, then, that f is Riemann integrable.

Moreover, f is equal to 0 on a set of full measure. So it follows that the Lebesgue integral of f on the interval $[0, 1]$ is 0. One may also calculate that the Riemann integral of f on $[0, 1]$ is equal to 0.

Example 4.3   Let

$$f(x) = \begin{cases} 0 & \text{if} & x \text{ is rational} \\ 1 & \text{if} & x \text{ is irrational}. \end{cases}$$

This function f is the Dirichlet function that we have seen before. It is discontinuous at every point of $[0, 1]$, so it is *not* Riemann integrable. In fact, one can see that, for a partition $\mathcal{P}$ with mesh arbitrarily fine, there are Riemann sums that are equal to 0 and there are Riemann sums that are equal to 1.

But this f is measurable and bounded, so it is certainly Lebesgue integrable. Furthermore,

$$\int_0^1 f(x)\,dx = 1.$$

Example 4.4   Let us revisit the example

$$\int_0^\infty \frac{\sin x}{x}\,dx. \qquad\qquad (\star)$$

From the point of view of the Riemann integral (more precisely, the *improper Riemann integral*), we think of this integral as the limit as $\epsilon \to 0^+$ of

$$\int_\epsilon^{1/\epsilon} \frac{\sin x}{x}\,dx.$$

Of course, $\sin x / x$ is bounded near the origin, so

$$\int_\epsilon^1 \frac{\sin x}{x}\,dx$$

converges.

And the integral

$$\int_1^{1/\epsilon} \frac{\sin x}{x}\,dx \equiv (*)$$

may be tamed with a simple integration by parts:

$$(*) = -\epsilon \cos(1/\epsilon) + \cos 1 + \int_1^{1/\epsilon} \frac{\cos x}{x^2}\,dx.$$

The integrand is bounded by $1/x^2$, so the limit of the integral as $\epsilon \to 0^+$ certainly exists.

We see, therefore, that the integral $(\star)$ exists as an improper Riemann integral.

Now let us examine $(\star)$ from the point of view of the Lebesgue integral. A function is Lebesgue integrable if and only if its positive part is integrable and its negative part is integrable. But the positive part of $\sin x / x$ exceeds

$$\sum_{j=1}^{\infty} \frac{\chi_{[\pi/4+2j\pi, 3\pi/4+2j\pi]}(x)}{x} \, .$$

It follows that the integral of this positive part exceeds

$$\sum_{j=1}^{\infty} \frac{\pi}{2} \cdot \frac{1}{3\pi/4 + 2j\pi} \, ,$$

and that diverges to $+\infty$. Similarly, the integral of the negative part diverges to $+\infty$.

We see then that the integral $(\star)$ converges as an improper Riemann integral, but not as a Lebesgue integral. As an improper Riemann integral, $(\star)$ is like a convergent alternating series. As a Lebesgue integral, it is more like the absolute value of that series.

Example 4.5   The integral

$$\int_0^{\infty} \frac{\sin x}{x^{3/2}} \, dx$$

converges both as an improper Riemann integral and as a Lebesgue integral. We leave the details as an exercise. We note that the integral of the absolute value of the integrand actually converges; no integrations by parts are either necessary or relevant.

EXERCISES

1. Suppose that $f_1 \geq f_2 \geq \cdots$ are continuous functions on $[0, 1]$, which converge pointwise to a continuous function f. May we conclude that the Riemann integrals of the f_j converge to the Riemann integral of f? May we conclude that the Lebesgue integrals of the f_j converge to the Lebesgue integral of f?

2. Let f_j be the characteristic function of the interval $[j/16, (j+1)/16]$, where all numbers are interpreted modulo 1 (that is, take the fractional part of each number). Discuss the limit of $\int_{\mathcal{R}} f_j(x) \, dx$, both from the point of view of the Riemann integral and from the point of view of the Lebesgue integral.

3. Let f be a Riemann integrable function on $[0, 1]$. Let $\epsilon > 0$. Show that there is a piecewise linear, continuous function φ on $[0, 1]$ such that

$$\int_0^1 |f(x) - \varphi(x)|\, dx < \epsilon,$$

where the integral here is the Riemann integral. Prove a similar statement for the Lebesgue integral.

4. Prove that any Lebesgue integrable function may be approximated in measure by a sequence of Riemann integrable functions.

5. Give an example of a function on the unit interval that is Lebesgue integrable, but which fails to be Riemann integrable because it is unbounded.

6. Give an example of a function on the unit interval that is Lebesgue integrable, but which fails to be Riemann integrable because it is discontinuous on a set with measure $1/2$.

CHAPTER 5

Other Theories of the Integral

In this chapter, we treat a number of alternative theories of the integral, or of measure. We provide few proofs, but instead concentrate on the exposition. The goal is to give a flavor of different ways to approach the integral concept.

5.1 THE DANIELL INTEGRAL

Named after Percy J. Daniell, the Daniell integral is a theory of integration that is equivalent to the Lebesgue integral but avoids the necessity of first developing measure theory. The Daniell integral is particularly useful in functional analysis contexts.

We begin by defining the *elementary functions*. This is a family $\mathcal{H}$ of functions on $\mathcal{R}$ that satisfies

1. $\mathcal{H}$ is a linear space equipped with the operations of addition and scalar multiplication.

2. If $h \in \mathcal{H}$ then $|h| \in \mathcal{H}$.

To each $h \in \mathcal{H}$, we assign a real number Ih, which we think of as the *elementary integral* of h. The elementary integral satisfies these three axioms:

(a) If $h, k \in \mathcal{H}$, and $\alpha, \beta \in \mathcal{R}$, then

$$I(\alpha h + \beta k) = \alpha Ih + \beta Ik \,.$$

(b) If $h(x) \geq 0$, then $Ih \geq 0$.

(c) If h_j are elementary functions with $h_1 \geq h_2 \geq h_3 \geq \cdots$ and $h_j \to 0$ pointwise, then $Ih_j \to 0$.

We say that a set S has *measure zero* if, for any $\epsilon > 0$, there is a nondecreasing sequence of nonnegative elementary functions h_j in $\mathcal{H}$ so that $\sup_j h_j(x) \geq 1$ for $x \in S$ and $Ih_j < \epsilon$ for every j. If a statement is true for all x outside a set of measure zero, then we say that the statement is true *almost everywhere*.

Let L^+ be the collection of functions f which are the limits of nondecreasing sequences h_j of elementary functions almost everywhere in such a manner that the set of integrals Ih_j is bounded. Then we set

$$If = \lim_{j \to \infty} Ih_j \,.$$

So defined, the class L^+ is not closed under subtraction and scalar multiplication by negative constants. But we may extend the class by considering functions $\varphi = f - g$ for $f, g \in L^+$. Then we set

$$I\varphi = If - Ig.$$

[This is similar to what we do with the Lebesgue integral.]

The nature of the Daniell integral that we have constructed here depends critically on the initial set of elementary functions. If we take the elementary functions to be step functions (i.e., functions that are piecewise constant on intervals), then the resulting measure is equivalent to Lebesgue measure. If we instead take the elementary functions to be the continuous functions, and use the Riemann integral to define the elementary integral, then Daniell's construction gives a measure that is equivalent to Lebesgue measure. As a third example, if we take the elementary functions to be generated by the functions $\sin jx$ for j an odd, positive integer, then the resulting class of Daniell integrable functions is distinctly smaller.

5.2   THE RIEMANN-STIELTJES INTEGRAL

For many purposes, such as integration by parts, it is natural to formulate the integral in a more general context than we have considered in the first two chapters. Our new formulation is called the *Riemann-Stieltjes integral* and is described below.

Fix an interval $[a, b]$ and a monotonically increasing function α on $[a, b]$. If $\mathcal{P} = \{p_0, p_1, \ldots, p_k\}$ is a partition of $[a, b]$, then let $\Delta\alpha_j = \alpha(p_j) - \alpha(p_{j-1})$. Let f be a bounded function on $[a, b]$ and define *the upper Riemann sum* of f with respect to α and the *lower Riemann sum* of f with respect to α as follows:

$$\mathcal{U}(f, \mathcal{P}, \alpha) = \sum_{j=1}^{k} M_j \Delta\alpha_j$$

and

$$\mathcal{L}(f, \mathcal{P}, \alpha) = \sum_{j=1}^{k} m_j \Delta\alpha_j.$$

Here the notation M_j denotes the supremum of f on the interval $I_j = [p_{j-1}, p_j]$ and m_j denotes the infimum of f on I_j.

In the special case $\alpha(x) = x$, the Riemann sums discussed here have a form similar to the Riemann sums considered in the first two chapters. Moreover,

$$\mathcal{L}(f, \mathcal{P}, \alpha) \le \mathcal{R}(f, \mathcal{P}) \le \mathcal{U}(f, \mathcal{P}, \alpha).$$

We define

$$I^*(f) = \inf \mathcal{U}(f, \mathcal{P}, \alpha)$$

and

$$I_*(f) = \sup \mathcal{L}(f, \mathcal{P}, \alpha).$$

Here the supremum and infimum are taken with respect to all partitions of the interval $[a, b]$. These are, respectively, the *upper* and *lower integrals* of f with respect to α on $[a, b]$.

By definition, it is always true that, for any partition $\mathcal{P}$,

$$\mathcal{L}(f, \mathcal{P}, \alpha) \leq I_*(f) \leq I^*(f) \leq \mathcal{U}(f, \mathcal{P}, \alpha).$$

It is natural to declare the integral to exist when the upper and lower integrals agree:

Definition 5.1   Let α be a monotone increasing function on the interval $[a, b]$, and let f be a bounded function on $[a, b]$. We say that the *Riemann-Stieltjes integral of f with respect to α* exists if

$$I^*(f) = I_*(f).$$

When the integral exists, we denote it by

$$\int_a^b f \, d\alpha.$$

Notice that the definition of Riemann-Stieltjes integral is different from the definition of Riemann integral that we used in the first two chapters. It turns out that, when $\alpha(x) = x$, the two definitions are equivalent (this assertion is explored in the exercises). In the present generality, it is easier to deal with upper and lower integrals in order to determine the existence of integrals.

Definition 5.2   Let $\mathcal{P}$ and $\mathcal{Q}$ be partitions of the interval $[a, b]$. If each point of $\mathcal{P}$ is also an element of $\mathcal{Q}$, then we call $\mathcal{Q}$ a *refinement* of $\mathcal{P}$.

Notice that the refinement $\mathcal{Q}$ is obtained by adding points to $\mathcal{P}$. The mesh of $\mathcal{Q}$ shall be less than or equal to that of $\mathcal{P}$. The following lemma enables us to deal effectively with our new language:

Lemma 5.3   *Let $\mathcal{P}$ be a partition of the interval $[a, b]$ and f a function on $[a, b]$. Fix a monotone increasing function α on $[a, b]$. If $\mathcal{Q}$ is a refinement of $\mathcal{P}$, then*

$$\mathcal{U}(f, \mathcal{Q}, \alpha) \leq \mathcal{U}(f, \mathcal{P}, \alpha)$$

and

$$\mathcal{L}(f, \mathcal{Q}, \alpha) \geq \mathcal{L}(f, \mathcal{P}, \alpha).$$

Proof. Since Q is a refinement of P, it holds that any interval I_ℓ arising from Q is contained in some interval $J_{j(\ell)}$ arising from P. Let M_{I_ℓ} be the supremum of f on I_ℓ and $M_{J_{j(\ell)}}$ the supremum of f on the interval $J_{j(\ell)}$. Then $M_{I_\ell} \leq M_{J_{j(\ell)}}$. We conclude that

$$\mathcal{U}(f, Q, \alpha) = \sum_\ell M_{I_\ell} \Delta\alpha_\ell \leq \sum_\ell M_{J_{j(\ell)}} \Delta\alpha_\ell .$$

We rewrite the right-hand side as

$$\sum_j M_{J_j} \left(\sum_{I_\ell \subseteq J_j} \Delta\alpha_\ell \right) .$$

However, because α is monotone, the inner sum simply equals $\alpha(p_j) - \alpha(p_{j-1}) = \Delta\alpha_j$. Thus, the last expression is equal to $\mathcal{U}(f, P, \alpha)$, as desired.

A similar argument applies to the lower sums. $\square$

Example 5.4 Let $[a, b] = [0, 10]$ and let $\alpha(x)$ be the *greatest integer function*.[1] That is, $\alpha(x)$ is the greatest integer that does not exceed x. So, for example, $\alpha(0.5) = 0$, $\alpha(2) = 2$, and $\alpha(-3/2) = -2$. Certainly α is a monotone increasing function on $[0, 10]$. Let f be any continuous function on $[0, 10]$. We shall determine whether

$$\int_0^{10} f \, d\alpha$$

exists and, if it does, calculate its value.

Let P be a partition of $[0, 10]$. By the lemma, it is to our advantage to assume that the mesh of P is smaller than 1. Observe that $\Delta\alpha_j$ equals the number of integers that lie in the interval I_j—that is, either 0 or 1. Let $I_{j_0}, I_{j_2}, \ldots I_{j_{10}}$ be, in sequence, the intervals from the partition which do, in fact, contain each distinct integer (the first of these contains 0, the second contains 1, and so on up to 10). Then

$$\mathcal{U}(f, P, \alpha) = \sum_{\ell=0}^{10} M_{j_\ell} \Delta\alpha_{j_\ell} = \sum_{\ell=1}^{10} M_{j_\ell}$$

and

$$\mathcal{L}(f, P, \alpha) = \sum_{\ell=0}^{10} m_{j_\ell} \Delta\alpha_{j_\ell} = \sum_{\ell=1}^{10} m_{j_\ell}$$

because any term in these sums corresponding to an interval not containing an integer must have $\Delta\alpha_j = 0$. Notice that $\Delta\alpha_{j_0} = 0$ since $\alpha(0) = \alpha(p_1) = 0$.

---

[1]In many texts, the greatest integer in x is denoted by $[x]$. We do not use that notation because it could get confused with our notation for a closed interval.

Let $\epsilon > 0$. Since f is uniformly continuous on $[0, 10]$, we may choose a $\delta > 0$ such that $|s - t| < \delta$ implies that $|f(s) - f(t)| < \epsilon/20$. If $m(\mathcal{P}) < \delta$, then it follows that $|f(\ell) - M_{j_\ell}| < \epsilon/20$ and $|f(\ell) - m_{j_\ell}| < \epsilon/20$ for $\ell = 0, 1, \ldots 10$. Therefore,

$$\mathcal{U}(f, \mathcal{P}, \alpha) < \sum_{\ell=1}^{10} \left(f(\ell) + \frac{\epsilon}{20} \right)$$

and

$$\mathcal{L}(f, \mathcal{P}, \alpha) > \sum_{\ell=1}^{10} \left(f(\ell) - \frac{\epsilon}{20} \right).$$

Rearranging the first of these inequalities leads to

$$\mathcal{U}(f, \mathcal{P}, \alpha) < \left(\sum_{\ell=1}^{10} f(\ell) \right) + \frac{\epsilon}{2}$$

and

$$\mathcal{L}(f, \mathcal{P}, \alpha) > \left(\sum_{\ell=1}^{10} f(\ell) \right) - \frac{\epsilon}{2}.$$

Thus, since $I_*(f)$ and $I^*(f)$ are trapped between $\mathcal{U}$ and $\mathcal{L}$, we conclude that

$$\left| I_*(f) - I^*(f) \right| < \epsilon.$$

We have seen that, if the partition is fine enough, then the upper and lower integrals of f with respect to α differ by at most ϵ. It follows that $\int_0^{10} f \, d\alpha$ exists. Moreover,

$$\left| I^*(f) - \sum_{\ell=1}^{10} f(\ell) \right| < \epsilon$$

and

$$\left| I_*(f) - \sum_{\ell=1}^{10} f(\ell) \right| < \epsilon.$$

We conclude that

$$\int_0^{10} f \, d\alpha = \sum_{\ell=1}^{10} f(\ell). \qquad \Box$$

The example demonstrates that the language of the Riemann-Stieltjes integral allows us to think of the integral as a generalization of the summation process. This is frequently useful, both philosophically and for practical reasons.

The next result, sometimes called Riemann's lemma, is crucial for proving the existence of Riemann-Stieltjes integrals.

Proposition 5.5   *Let α be a monotone increasing function on $[a, b]$ and f a bounded function on the interval. The Riemann-Stieltjes integral of f with respect to α exists if and only if, for every $\epsilon > 0$, there is a partition $\mathcal{P}$ such that*

$$|\mathcal{U}(f, \mathcal{P}, \alpha) - \mathcal{L}(f, \mathcal{P}, \alpha)| < \epsilon. \tag{$*$}$$

Proof. First assume that $(*)$ holds. Fix $\epsilon > 0$. Since $\mathcal{L} \leq I_* \leq I^* \leq \mathcal{U}$, inequality $(*)$ implies that

$$\left|I^*(f) - I_*(f)\right| < \epsilon.$$

But this means that $\int_a^b f \, d\alpha$ exists.

Conversely, assume that the integral exists. Fix $\epsilon > 0$. Choose a partition $\mathcal{Q}_1$ such that

$$\left|\mathcal{U}(f, \mathcal{Q}_1, \alpha) - I^*(f)\right| < \epsilon/2.$$

Likewise, choose a partition $\mathcal{Q}_2$ such that

$$|\mathcal{L}(f, \mathcal{Q}_2, \alpha) - I_*(f)| < \epsilon/2.$$

Since $I_*(f) = I^*(f)$, it follows that

$$|\mathcal{U}(f, \mathcal{Q}_1, \alpha) - \mathcal{L}(f, \mathcal{Q}_2, \alpha)| < \epsilon. \tag{$**$}$$

Let $\mathcal{P}$ be the common refinement of $\mathcal{Q}_1$ and $\mathcal{Q}_2$. Then we have, again by Lemma 5.3, that

$$\mathcal{L}(f, \mathcal{Q}_2, \alpha) \leq \mathcal{L}(f, \mathcal{P}, \alpha) \leq \int_a^b f \, d\alpha \leq \mathcal{U}(f, \mathcal{P}, \alpha) \leq \mathcal{U}(f, \mathcal{Q}_1, \alpha).$$

But, by $(**)$, the expressions on the far left and on the far right of these inequalities differ by less than ϵ. Thus, $\mathcal{P}$ satisfies the condition $(*)$. $\square$

We note in passing that the basic properties of the Riemann integral noted in Chapter 2 hold without change for the Riemann-Stieltjes integral. The proofs are left as exercises for you.

5.3 THE HENSTOCK-KURZWEIL INTEGRAL

Also known as the Denjoy integral and the Perron integral, this is a generalization of the Riemann integral. Part of the interest is in integrating functions like

$$f(x) = \frac{1}{x} \cdot \sin\left(\frac{1}{x^3}\right).$$

This function is *not* Lebesgue integrable.

The main difference between the Henstock-Kurzweil integral and the Riemann integral is that, in the former, we measure the fineness of a partition in a more general way. Namely, we have a function $\delta : [a, b] \to (0, \infty)$, which we call a *gauge*. If $a = p_0 < p_1 < p_2 < \cdots < p_k = b$ is a partition of $[a, b]$ and $t_j \in [p_{j-1}, p_j]$, then we require that $p_j - p_{j-1} < \delta(t_j)$. Such a partition is called δ-*fine*. Finally, Riemann sums $\mathcal{R}$ are formed as usual, and we say that I is the Henstock-Kurzweil integral of f on $[a, b]$ if, for every $\epsilon > 0$, there is a gauge δ such that whenever $\mathcal{P}$ is a partition that is δ-fine then

$$|\mathcal{R} - I| < \epsilon.$$

It is known, thanks to Cousin, that every gauge δ corresponds to a δ-fine partition $\mathcal{P}$. So the condition is not vacuous.

It is also known that a bounded function is Henstock-Kurzweil integrable if and only if it is Lebesgue integrable. In effect, the Henstock-Kurzweil integral is a non-absolutely convergent version of the Lebesgue integral.

5.4 HAUSDORFF MEASURE

The idea of Hausdorff measure does not lead to a new idea of integral, but it is related to the idea of integral and well worthy of our consideration. In this section only, we depart from custom and work in any dimensional Euclidean space $\mathcal{R}^N$.

First we consider Carathéodory's construction. Let $S \subseteq \mathcal{R}$ and $\delta > 0$. Let $\mathcal{U} = \{U_\alpha\}_{\alpha \in A}$ be an open covering of S. Call $\mathcal{U}$ a δ-admissible covering if each U_α is an open Euclidean N-interval of radius $0 < r_\alpha < \delta$. If $0 \le k \in \mathcal{Z}$, let M_k be the usual k-dimensional Lebesgue measure of the unit interval in $\mathcal{R}^k$ (e.g., $M_1 = 2$, $M_2 = \pi$, $M_3 = 4\pi/3$, etc.). Define

$$\mathcal{H}^k_\delta(S) = \inf\left\{\sum_{\alpha \in A} M_k r_\alpha^k : \mathcal{U} = \{U_\alpha\}_{\alpha \in A} \text{ is a } \delta - \text{admissible covering}\right\}.$$

Clearly, $\mathcal{H}^k_\delta(S) \le \mathcal{H}^k_{\delta'}(S)$ if $0 < \delta' < \delta$. Therefore, $\lim_{\delta \to 0} \mathcal{H}^k_\delta(S)$ exists in the extended real number system. The limit is called the *k-dimensional Hausdorff measure* of S and is denoted by $\mathcal{H}^k(S)$. The function $\mathcal{H}^k$ is an outer measure.

EXERCISES FOR THE READER

1. If $I \subseteq \mathcal{R}$ is a line segment, then $\mathcal{H}^1(I)$ is the usual Euclidean length of I. Also, $\mathcal{H}^0(I) = \infty$ and $\mathcal{H}^k(I) = 0$ for all $k > 1$.

2. If $S \subseteq \mathcal{R}$ is Borel, then $\mathcal{H}^N(S) = m(S)$, where m is Lebesgue N-dimensional measure.

3. If $S \subseteq \mathcal{R}$ is a discrete set, then $\mathcal{H}^0(S)$ is the number of elements of S.

4. Let $\alpha > 0$ and define $M_\alpha = \Gamma(1/2)^\alpha / \Gamma(1 + \alpha/2)$, (note that this is consistent with the preceding definition of M_k). Then define H^α for any $\alpha > 0$ by using Carathéodory's construction. Let $S \subseteq \mathcal{R}$ be the Cantor set. Compute $\alpha_0 = \sup\{\alpha > 0 : \mathcal{H}^\alpha(S) = \infty\}$. Also compute $\alpha_1 = \inf\{\alpha > 0 : \mathcal{H}^\alpha(S) = 0\}$. Then $\alpha_0 = \alpha_1$. What is its specific value? This number is called the *Hausdorff dimension* of S. What is the Hausdorff dimension of a regularly imbedded, k-dimensional, C^1 manifold in $\mathcal{R}^N$? It is in fact a deep theorem (see [FED]) that any rectifiable set $S \subseteq \mathcal{R}$ has the property that $\alpha_0 = \alpha_1$.

5.5   HAAR MEASURE

We take for granted that in $\mathcal{R}$ the most commonly used measure V (Lebesgue volume) has the property that $V(A) = V(b + A)$ for any measurable set A and any translation by an element $b \in \mathcal{R}$. In fact, this translation invariance essentially characterizes Lebesgue measure on $\mathcal{R}$. However, consider instead the space $\mathcal{R}^+ \equiv \{x \in \mathcal{R} : x > 0\}$ with the group operation being multiplication (instead of addition). Now what is the invariant measure?

In fact, the reader may verify that the measure dx/x is invariant under the group action. Indeed, if A is a measurable set and $b \in \mathcal{R}^+$, then

$$\int_{\mathcal{R}^+} \chi_A(x \cdot b) \frac{dx}{x} = \int_{\mathcal{R}^+} \chi_A(x) \frac{dx}{x}.$$

More generally, is it possible to find an invariant measure on any topological group? By a topological group here we mean a topological space that also comes equipped with a binary operation that induces a group structure on the underlying set. We require that the group operations (product and inverse) be continuous in the given topology. Examples of topological groups are

$$(\mathcal{R}, +), \ (\mathcal{T}, \cdot), \ (O(N), \cdot), \ (SO(N), \cdot).$$

While an invariant measure, called *Haar measure*, exists on any locally compact group, we shall concentrate our efforts in the present section on *compact groups*. One advantage of compact groups is that the left-invariant Haar measure and the right-invariant Haar measure are identical. For our purposes, the study of compact groups will suffice.

5.5.1   THE FUNDAMENTAL THEOREM

The basic theorem about the existence and uniqueness of Haar measure is as follows.

Theorem 5.6   *Let G be a compact topological group. There is a unique, invariant integral λ on G such that $\lambda(1) = 1$. The invariance of λ means that, if φ is a continuous function on G, if $g \in G$, and if $\varphi_g(x) \equiv \varphi(gx)$, then*

$$\lambda(\varphi) = \lambda(\varphi_g).$$

Corollary 5.7   *Let G be a compact topological group. There is a unique, invariant Radon measure μ on G such that $\mu(G) = 1$. The invariance of μ means that, for all sets $A \subseteq G$ and $g \in G$,*

$$\mu(A) = \mu\{ga : a \in A\} = \mu\{ag^{-1} : a \in A\}.$$

If G is a compact topological group and also happens to be a metric space (such as the orthogonal group—see below), then we say that the metric d is *invariant* if

$$d(gh, gk) = d(hg, kg) = d(h, k)$$

for any g, h, k in the group. It follows, for such a metric, that $gI(h, r) = I(gh, r)$ for any (open) metric interval I. Since the Haar measure is invariant, we conclude that the Haar measure of all intervals with the same radii are the same. In fact, this property characterizes Haar measure, as we shall now see.

Definition 5.8   A Borel regular measure μ on a metric space X is called *uniformly distributed* if the measures of all non-trivial intervals are positive and, in addition,

$$\mu(I(x, r)) = \mu(I(y, r)) \quad \text{for all } x, y \in X, 0 < r < \infty.$$

Proposition 5.9   *Let μ and ν be uniformly distributed, Borel regular measures on a separable metric space X. Then there is a positive constant c such that $\mu = c \cdot \nu$.*

It is a matter of some interest to determine the Haar measure on some specific groups and symmetric spaces. We note that Haar measure on $\mathcal{R}$ is Lebesgue measure (or any constant multiple thereof). Since this group is non-compact, we must forego the stipulation that the total mass of the measure be 1.

In this book, we are particularly interested in groups that bear on the geometry of Euclidean space. We have already noted the Haar measure on the multiplicative reals, which corresponds to the dilation group. And the preceding paragraph treats the Haar measure of the group of translations.

EXERCISES

1. Verify that, when $\alpha(x) = x$, then the Riemann-Stieltjes integral of a function f with respect to α on $[a, b]$ is just the same as the Riemann integral of f on $[a, b]$.

2. Define $\alpha(x)$ by the condition that $\alpha(x) = -x + k$ when $k \le x < k + 1$. Calculate

$$\int_2^7 t^2 \, d\alpha(t) .$$

3. Let $[x]$ be the greatest integer function as discussed in the text. Define the "fractional part" function by the formula $\alpha(x) = x - [x]$. Explain why this function has the name "fractional part." Calculate

$$\int_0^5 x \, d\alpha .$$

4. Let β be a monotone increasing function on the interval $[a, b]$. Set $m = \beta(a)$ and $M = \beta(b)$. For any number λ lying between m and M set $S_\lambda = \{x \in [a, b] : \beta(x) > \lambda\}$. Prove that S_λ must be an interval. Let $\ell(\lambda)$ be the length of S_λ. Then prove that

$$\int_a^b \beta(t)^p \, dt = -\int_m^M s^p d\ell(s)$$
$$= \int_0^M \ell(s) \cdot p \cdot s^{p-1} ds.$$

6. Find a natural way to express integration by parts in the language of the Riemann-Stieltjes integral. How would one prove this new formula?

7. Prove that any function that is Daniell integrable is also Lebesgue integrable.

8. Calculate Haar measure on the group which is the collection of positive real numbers equipped with the binary operation of multiplication of reals.

10. The Hausdorff dimension of a set S is the infimum of all $\alpha > 0$ such that $H^\alpha(S) = 0$. For any $0 < \lambda < 1$, imitate the standard construction of the Cantor ternary set (see [KRA1]) to produce a subset of the reals that has Hausdorff dimension λ.

11. The Hausdorff dimension of a set S is the infimum of all $\alpha > 0$ such that $H^\alpha(S) = 0$. What is the Hausdorff dimension of the set of rational numbers? What is the Hausdorff dimension of the set of irrational numbers?

12. The Hausdorff dimension of a set S is the infimum of all $\alpha > 0$ such that $H^\alpha(S) = 0$. Show that any nonempty open set in $\mathcal{R}$ has Hausdorff dimension 1.

13. Prove that the Riemann-Stieltjes integral is linear—that is, that is respects addition and scalar multiplication. Prove that

$$\int_a^b f \, d\alpha = \int_a^c f \, d\alpha + \int_c^b f \, d\alpha \, .$$

14. Prove that if $f \leq g$ then, in the sense of the Riemann-Stieltjes integral,

$$\int f \, d\alpha \leq \int g \, d\alpha \, .$$

Bibliography

[BRO] A. B. Brown, A proof of the Lebesgue condition for Riemann integrability, *Am. Math. Monthly* 43(1936), 396–398. DOI: 10.2307/2301737 12

[FED] H. Federer, *Geometric Measure Theory*, Springer-Verlag, New York, 1969. 82

[FOL] G. B. Folland, *Real Analysis: Modern Techniques and Their Applications*, John Wiley & Sons, New York, 1984. 53, 54, 66

[HUW] W. Hurewicz and H. Wallman, *Dimensional Theory*, Oxford University Press, London, 1941. 56

[JEC] T. J. Jech, *The Axiom of Choice*, North-Holland, Amsterdam, 1973. 36

[KRA1] S. G. Krantz, *Real Analysis and Foundations*, CRC Press, Boca Raton, FL, 1992. 4, 58, 84

[KRA2] S. G. Krantz, *A Panaroma of Harmonic Analysis*, Mathematical Association of America, Washington, D.C., 1999. 59

[KRP] S. G. Krantz and H. R. Parks, *The Geometry of Domains in Space*, Birkhäuser Publishing, Boston, MA, 1996. 56

[MAT] P. Mattila, *Geometry of Sets and Measures in Euclidean Spaces : Fractals and Rectifiability*, Cambridge University Press, Cambridge, 1995. 62

[RIE] B. Riemann, Über die Darstellbarkeit einer Function durch eine trigonometrische Reihe, Habilitationsschrift, 1854, *Abhandlungen der Königlichen Gesellschaft der Wissenschaften zu Göttingen* 13(1868). 1

[RIT] J. F. Ritt, *Integration in Finite Terms*, Columbia University Press, New York, 1948. 16

[ROY] H. Royden, *Real Analysis*, 3rd ed., Macmillan, New York, 1988. 3

[RUD] W. Rudin, *Real and Complex Analysis*, 3rd ed., McGraw-Hill, New York, 1987. 53

[STE] E. M. Stein, *Harmonic Analysis: Real-variable Methods, Orthogonality, and Oscillatory Integrals*, Princeton University Press, Princeton, N.J., 1993. 59

Author's Biography

STEVEN G. KRANTZ

Steven G. Krantz was born in San Francisco, California in 1951. He received the B.A. degree from the University of California at Santa Cruz in 1971 and the Ph.D. from Princeton University in 1974.

Krantz has taught at UCLA, Princeton University, Penn State, and Washington University in St. Louis. He has served as Chair of the latter department.

Krantz has directed 9 Masters theses and 18 Ph.D. theses. He has been awarded the UCLA Alumni Foundation Distinguished Teaching Award, the Chauvenet Prize, and the Beckenbach Book Prize.

Krantz has published over 155 scholarly articles and over 55 books. He is currently the Editor of the Notices of the American Mathematical Society.

Index

Printed in the United States
by Baker & Taylor Publisher Services